T0133357

Ultrafast photophysics and photochemistry

after excited state intramolecular charge transfer

in the liquid phase

Zur Erlangung des akademischen Grades eines

DOKTORS DER NATURWISSENSCHAFTEN

(Dr. rer. nat.)

Fakultät für Chemie und Biowissenschaften

Karlsruher Institut für Technologie (KIT) – Universitätsbereich

genehmigte

DISSERTATION

von

Hanna A. Ernst

aus

Bühl (Baden)

Dekan: Prof. Dr. Peter Roesky

Referent: PD Dr. Andreas-Neil Unterreiner

Korreferent: Prof. Dr. Manfred Kappes

Tag der mündlichen Prüfung: 17.04.2015

Bibliografische Information der Deutschen Nationalbibliothek

Die Deutsche Nationalbibliothek verzeichnet diese Publikation in der
Deutschen Nationalbibliografie; detaillierte bibliografische Daten sind
im Internet über http://dnb.d-nb.de abrufbar.

ISBN 978-3-8325-3984-9

Logos Verlag Berlin GmbH
Comeniushof, Gubener Str. 47,
10243 Berlin
Tel.: +49 (0)30 42 85 10 90
Fax: +49 (0)30 42 85 10 92
INTERNET: http://www.logos-verlag.de

For Bastian

For my sister Lena

For my friends Simone, Tristan and Mario

Hiermit erkläre ich, dass ich diese Arbeit selbständig angefertigt habe und keine anderen als die angegebenen Hilfsmittel und Quellen verwendet wurden.

Summary

In the present work, the ultrafast excited state dynamics after intramolecular charge transfer excitation in the ultraviolet (UV) of selected organic compounds have been studied by means of steady state UV-Vis and femtosecond time-resolved transient absorption spectroscopy in the liquid phase. The investigated systems comprise small donor-acceptor molecules as well as type-I photoinitiators. The experimental findings were assisted by ab initio calculations at the (TD)-DFT and the CASSCF level of theory. In the following, a brief overview of the characteristic relaxation features is provided.

Ultrafast dynamics of nitrophenols and corresponding nitrophenolates

The small donor-acceptor compounds *ortho-*, *meta-* and *para*-nitrophenol are characterized by intramolecular charge transfer in the ultraviolet. The induced electron redistribution from the hydroxyl to the nitro group can be associated with a transition into the first bright excited singlet state. It has been found that ultrafast intersystem crossing on sub-picosecond time scale is common for all isomers. Their further relaxation behavior, however, strongly depends on the different isomerism.

For *ortho*-nitrophenol, the experimental as well as theoretical findings indicate an excited state intramolecular proton transfer from the hydroxyl to the adjacent nitro group, and thus transient formation of the tautomeric *aci*-nitro form triggered by the strong charge transfer character of the initially excited state. The proton transfer itself is found to be accompanied by an out-of-plane twist of the newly formed nitrous acid (HONO) group. These structural changes lead to an expeditious drop of the energy gap between the initially excited and the ground state S_0, finally resulting in a conical intersection between both states. Return to S_0 was monitored by stimulated emission induced by the probe pulse within 0.2-0.3 picoseconds nearly independent of the employed solvents.

On the contrary, the dynamics of the *meta* and *para* isomer reveal a strong solvent dependence. Photodegradation culminating in the release of nitrogen(II)oxide (NO·) and the related aryloxy radical (Ar–O·) is found in chloroform and 2-propanol within 6-123 picoseconds. Thereby, the radical generation most likely originates from the upper unrelaxed triplet manifold. In water, photoreaction plays only a minor role for *para*-nitrophenol and does not occur for the *meta* form due to efficient coupling and energy transfer to surrounding solvent molecules. Thus, for the *meta* and *para* isomer, population of triplet states via ultrafast intersystem crossing is suggested to be the main relaxation pathway in water. Moreover, formation of the corresponding nitrophenolates becomes important in this case. As a consequence, superposition of transient absorption features of both the protonated and deprotonated forms were observed.

The relaxation behavior of the latter reveals a strong analogy to the processes observed for *ortho*-nitrophenol. All nitrophenolates exhibit an intramolecular charge transfer in their initially excited singlet state involving the oxygen atom, directly linked to the aromatic frame, and the nitro group. As for *ortho*-nitrophenol, return to the electronic ground state was observed by stimulated emission within 0.2-6 picoseconds. However, changes of the excited state energy solely evolve by out-of-plane rotation of the nitro group until a conical intersection with S_0 is reached.

Ultrafast dynamics of selected type-I photoinitiators

The dynamics of the three type-I photoinitiators benzoin (2-hydroxy-1,2-diphenylethanone, Bz), 4-methyl-benzoin (4MB) and 2-methyl-4'-(methylthio)-2-morpholinopropiophenone (MMMP) were studied in methylisobutyrate. After intramolecular charge transfer in the ultraviolet, these initiators decompose into their radical fragments due to α-cleavage in the first excited triplet state. The excitation itself is characterized by electron redistribution from the hydroxybenzyl and morpholino moiety, respectively, to the benzoyl fragment. Comparing the photoinduced dynamics of Bz and 4MB, similar relaxation behavior is observed and time constants of 50-80 and 80-180 picoseconds are found for radical formation.

In contrast, significantly higher complexity of the relaxation pathways is discovered for MMMP resulting in an increased time constant for radical generation of >1800 picoseconds. Moreover, a quantitative analysis of the underlying relaxation mechanisms is proposed suggesting higher capability of radical formation for MMMP and 4MB compared to Bz.

A quantitative correlation, however, remains difficult with respect to initiation efficiencies for polymerization of methyl methacrylate derived from pulsed laser polymerization experiments. The combination of transient absorption spectroscopy and post-mortem analysis of the produced polymer has demonstrated that additional information is required for a quantitative evaluation of initiation efficiencies. For instance, the fragments' reactivity towards vinyl bonds cannot be neglected, when comparing different radical fragments originating from disparate source molecules.

Zusammenfassung

Im Rahmen der vorliegenden Arbeit wurde die ultraschnelle Dynamik elektronisch angeregter Zustände, welche durch Ladungstransfer-Anregung im ultravioletten (UV) Bereich gekennzeichnet sind, mit Hilfe von stationärer UV-Vis- und zeitaufgelöster transienter Absorptions-Spektroskopie auf der Femtosekunden-Zeitskala untersucht. Die ausgewählten organischen Systeme bestanden zum einen aus kleinen aromatischen Donor-Akzeptor-Molekülen und zum anderen aus Typ-I-Photoinitiatoren. Für die Interpretation der experimentellen Befunde wurden ferner ab initio Rechnungen auf (TD)-DFT- und CASSCF-Basis durchgeführt. Die photoinduzierte Dynamik der untersuchten Verbindungen ist im Folgenden beschrieben.

Ultraschnelle Dynamik von Nitrophenolen und ihren korrespondierenden Nitrophenolaten

Die Donor-Akzeptor-Moleküle *ortho-*, *meta-* und *para-*Nitrophenol zeichnen sich durch intramolekulare Ladungstransfer-Anregungen im UV-Bereich aus. Die Anregung in den ersten hellen Singulett-Zustand ist durch eine Ladungsverschiebung zwischen der Hydroxyl- und der Nitrogruppe geprägt. Als charakteristischer Relaxationskanal aller Isomere konnte die Population von Triplett-Zuständen über Singulett-Triplett-Interkombination (Intersystem Crossing) auf der sub-Pikosekunden-Zeitskala identifiziert werden. In ihren weiteren dynamischen Prozessen unterscheiden sich die Isomere allerdings sehr stark in Abhängigkeit des jeweils vorliegenden Substitutionsmusters.

Für *ortho-*Nitrophenol zeigen die experimentellen wie auch die theoretischen Befunde einen intramolekularen Protonentransfer zwischen der Hydroxyl- und der benachbarten Nitrogruppe. Dieser wird durch den Ladungstransfer-Charakter des angeregten Singulett-

Zustands induziert und führt zur Bildung der instabilen tautomeren Form von *ortho*-Nitrophenol. Der Protonentransfer selbst geht mit einer Drehung der neu gebildeten HONO-Gruppe (salpetrige Säureeinheit) aus der aromatischen Ebene einher. Diese strukturellen Änderungen in *ortho*-Nitrophenol bewirken eine extreme Verringerung des energetischen Abstands zwischen den Potentialflächen des angeregten Singulett- und des elektronischen Grundzustands und führen schließlich zu einer konischen Überschneidung zwischen beiden Zuständen. Die Rückkehr angeregter Moleküle in den Grundzustand konnte durch instantan auftretende, stimulierte Emission verfolgt werden, die durch den Abfragepuls induziert wurde. Sowohl das Auftreten als auch das Abklingen der stimulierten Emission innerhalb von 0,2-0,3 Pikosekunden wurde unabhängig von den verwendeten Lösungsmitteln beobachtet.

Für *meta*- und *para*-Nitrophenol hingegen wurde eine starke Lösungsmittelabhängigkeit des Relaxationsverhaltens gefunden. Die photoinduzierte Bildung von Stickstoff(II)oxid (NO·) und den zugehörigen Aryloxy-Radikalen (Ar–O·) wurde in Chloroform und 2-Propanol auf einer Zeitskala von 6-123 Pikosekunden nachgewiesen. Die Entstehung der Radikale findet dabei höchst wahrscheinlich nach Intersystem Crossing aus angeregten Triplett-Zuständen statt. Im Gegensatz dazu spielt diese Zersetzung in Wasser für *para*-Nitrophenol nur eine untergeordnete Rolle und konnte für *meta*-Nitrophenol überhaupt nicht beobachtet werden. Dieser Befund lässt sich zurückführen auf eine stärkere Wechselwirkungen zwischen angeregten und umgebenden Lösungsmittelmolekülen, und damit auf eine effizientere Energieübertragung in polaren Lösungsmitteln. Der dominierende Relaxationskanal in Wasser ist durch Intersystem Crossing und die nachfolgende Relaxation innerhalb von Triplett-Zuständen charakterisiert. Weiterhin konnte gezeigt werden, dass bei der Verwendung von Wasser als Lösungsmittel auch die Bildung der entsprechenden Nitrophenolate berücksichtigt werden muss. Die Änderung der transienten Absorption setzt sich dabei als Überlagerung der protonierten und deprotonierten Spezies zusammen.

Das Relaxationsverhalten der deprotonierten Formen weist eine hohe Analogie zu den beobachteten dynamischen Prozessen bei *ortho*-Nitrophenol auf. Für alle Nitrophenolate kann die Anregung in den ersten hellen Singulett-Zustand durch intramolekularen Ladungstransfer zwischen dem Sauerstoffatom des Phenolats und der Nitrogruppe beschrieben werden. Die Rückkehr angeregter Moleküle in ihren elektronischen Grundzustand wurde, wie im Fall von *ortho*-Nitrophenol, durch stimulierte Emission auf einer Zeitskala von 0,2-6 Pikosekunden beobachtet. Änderungen in der Energie des ersten angeregten Zustands lassen sich hier jedoch ausschließlich auf eine Drehung der Nitrogruppe aus der aromatischen Ringebene zurückführen. Diese Torsion führt letztlich zu einer konischen Überschneidung mit dem Grundzustand.

Ultraschnelle Dynamik ausgewählter Typ-I-Photoinitiatoren

Die photoinduzierte Dynamik der drei Typ-I-Photoinitiatoren Benzoin (2-Hydroxy-1,2-diphenylethanon, Bz), 4-Methyl-Benzoin (4MB) sowie 2-Methyl-4'-(methylthio)-2-morpholinopropiophenon (MMMP) wurden im Lösungsmittel Methylisobutyrat untersucht. Die ausgewählten Photoinitiatoren zeichnen sich durch intramolekularen Ladungstransfer im UV-Bereich und anschließender Radikalbildung durch α-Spaltung im ersten angeregten Triplett-Zustand aus. Der auftretende Ladungs-Transfer findet zwischen der Hydroxybenzyl-bzw. der Morpholino-Gruppe und dem Benzoyl-Fragment statt. Ein ähnliches Relaxationsverhalten wurde für Bz und 4MB beobachtet. Die ermittelten Zeitkonstanten für die Radikalbildung liegen in einem Bereich von 50-80 bzw. 80-180 Pikosekunden.

Im Gegensatz hierzu weist MMMP eine wesentlich komplexere Dynamik auf. Für die Radikalbildung konnte eine Zeitkonstante von >1800 Pikosekunden ermittelt werden. Weiterhin wurde eine quantitative Analyse der zugrunde liegenden Relaxationsmechanismen durchgeführt, um die Fähigkeit der Photoinitiatoren hinsichtlich der α-Spaltung zu untersuchen. Dabei wurde eine höhere Tendenz zur Radikalbildung für MMMP und 4MB im Vergleich zu Bz gefunden. Verglichen mit relativen Einbauverhältnissen der Initiatoren, welche aus gepulsten Polymerisationsexperimenten in Methylmethacrylat abgeleitet wurden, bleibt eine quantitative Korrelation dennoch schwierig. Mit der Verknüpfung von transienter Absorptions-Spektroskopie und der post-mortem Analyse der gebildeten Polymere konnte gezeigt werden, dass zusätzliche Informationen für eine quantitative Analyse der Initiierungseigenschaften erforderlich sind. Zum Beispiel kann bei der Untersuchung von verschiedenen Radikalfragmenten, die von unterschiedlichen Radikalvorläufern gebildet werden, die Reaktivität der Fragmente gegenüber Vinyl-Bindungen nicht vernachlässigt werden.

Contents

List of abbreviations

Ar·	aryl radical
Ar–O·	aryloxy radical
a.u.	arbitrary units
B	radical fragment according to figure 5.2
BBO	β-barium borate
BiBuQ	(4,4'''-Bis-(2-butyloctyloxy)-p-quaterphenyl)
Bn	radical fragment according to figure 5.2
BS	beam splitter
Bz	benzoin (2-hydroxy-1,2-diphenylethanone)
CAS	complete active space
CASSCF	complete active space self-consistent field
CI	conical intersection
CPA 2210	employed laser system
CT	charge transfer
DADS	decay associated difference spectra
DBU	1,8-diazabicyclo[5.4.0]undec-7-ene
DETC	7-diethylamino-3-thenoylcoumarin
DFM	difference frequency mixing
DFT	density functional theory
DM	dielectric mirror
ΔOD	change of optical density
ESA	excited state absorption
ESIPT	excited state intramolecular proton transfer

ε	extinction coefficient
FM	focusing mirror
fs	femtosecond
FWHM	full width of half maximum
GB	glass block
GD	group delay
GDD	group delay dispersion
GVM	group velocity mismatch
HFSCF	Hatree-Fock self-consistent field
HGSA	hot ground state absorption
HM	half-mirror
HOMO	highest occupied molecular orbital
HONO	nitrous acid
IC	internal conversion
ICT	intramolecular charge transfer
ISC	intersystem crossing
KTP	potassium titanyl phosphate ($KTiOPO_4$)
LUMO	lowest unoccupied molecular orbital
λ_{probe}	probe wavelength
λ_{pump}	pump wavelength
MB	radical fragment according to figure 5.2
4MB	4-methyl-benzoin
MIB	methylisobutyrate
Me	mesitil
MECI	minimum energy conical intersection
MEP	minimum energy pathway
MMMP	2-methyl-4'-(methylthio)-2-morpholinopropiophenone
m-NP	*meta*-nitrophenol
m-NP$^{\ominus}$	*meta*-nitrophenolate
μs	microsecond
N	radical fragment according to figure 5.2

NIR	near infrared
NO·	nitrogen(II)oxide radical
NO_2·	nitrogen(IV)oxide radical
NOPA	non-collinear optical parametric amplifier
NP(s)	nitrophenol(s)
NPAH(s)	nitrated polycyclic aromatic hydrocarbon(s)
ns	nanosecond
OD	optical density
o-NP	*ortho*-nitrophenol
o-NP$^\ominus$	*ortho*-nitrophenolate
OPA	optical parametric amplification
PD	photodiode
PES	potential energy surface
PK	prism compressor
PLP-ESI-MS	pulsed laser polymerization combined with subsequent ionization mass spectrometry
p-NP	*para*-nitrophenol
p-NP$^\ominus$	*para*-nitrophenolate
ps	picosecond
S_0	electronic singlet ground state
S_n	nth excited singlet state
SE	stimulated emission
SFM	sum frequency mixing
SHG	second harmonic generation
SPM	self-phase-modulation
T	radical fragment according to figure 5.2
T_1	lowest excited triplet state
T_n	nth excited triplet state
TA	transient absorption
TA (sw)	transient absorption (single wavelength)
TA (wl)	transient absorption (white light)

TD-DFT	time-dependent density functional theory
THG	third harmonic generation
Ti:Sa	titanium doped sapphire
TMB	2,4,6-trimethyl-benzoin
UV	ultraviolet
Vis	visible
VOC(s)	volatile organic compound(s)
VR	vibrational relaxation
WL	white light

1 Introduction

The invention of lasers and the subsequent development of new experimental spectroscopic techniques have decisively affected not only physical but also chemical and biological research.[1] The realization of the first ruby laser by Maiman in 1960[2] marked the beginning of an intensive advancement of ultrashort laser pulse generation. In 1961, pulse durations of a few nanoseconds (ns, 10^{-9} s) were produced by means of Q-switching (Hellwarth *et al.*)[3], closely followed by picosecond (ps, 10^{-12} s) laser pulses created by mode-locking in 1966 (De Maria *et al.*).[4, 5] Almost twenty years later, pulses characterized by a temporal width of six femtoseconds (fs, 10^{-15} s) were achieved.[6] With the development of femtosecond laser pulses from highly stable solid state Ti:sapphire lasers in 1991 the generation as well as the use of such ultrashort pulses became a standard laboratory tool.[7-12] Thereby, various spectroscopic applications are provided, since the tunability of a femtosecond laser pulse with respect to its central wavelength can be controlled by optical parametric amplification processes (section 2.1.4).[13-15] Pulse durations of few femtoseconds allow for the study of molecular dynamics, and thus for the investigation of fundamental processes in chemistry. In this research field, especially the so-called pump-probe technique has been established as a powerful tool.[1, 16]

One famous example of femtochemistry, which fascinates experimentalists and theoreticians until today, is the *cis-trans* isomerization of the 11-*cis*-retinal chromophore in the coupled receptor rhodopsin.[17] The conversion of 11-*cis*-retinal to its all-*trans* form has long been identified as the primary step in vision activated by photon absorption and culminating in stimulation of the optic nerve.[18, 19] Understanding the dynamics of the underlying isomerization, however, has been a challenging task for research over a long period of time. Only the invention of femtosecond-resolved pump-probe techniques made it possible to unravel the decisive relaxation step. Several studies determined the completeness of the photoisomerization within 200 fs[20-24] resulting in a high reaction quantum yield and efficient energy storage which affords the photochemical reaction cascade of rhodopsin.[21, 25, 26] This unique reactivity of rhodopsin is assigned to a conical intersection between the initially excited and the ground state along the isomerization coordinate.[17, 27, 28] Relaxation via conical intersection represents only one possibility of how molecules can relax after photoexcitation. In the following, a brief overview of different relaxation processes and the time scales on which they occur is provided.[29, 30]

The best way for discussion is to use a Jabłoński-type diagram as shown in figure 1.1.[31] Generally, the electronic excitation of molecules in their singlet ground state (S_0) caused by absorption of photons leads to transition into either the first (S_1) or a higher excited singlet state (S_n). Subsequently, depopulation of the initially excited state can occur via radiationless internal conversion (IC) into a lower excited singlet state or into highly excited vibrational states of the electronic ground state. Furthermore, a second non-radiative relaxation pathway is given by intersystem crossing (ISC) into (in most cases) the upper triplet manifold (T_n), followed by IC into the first excited triplet state (T_1) and finally ISC into S_0.

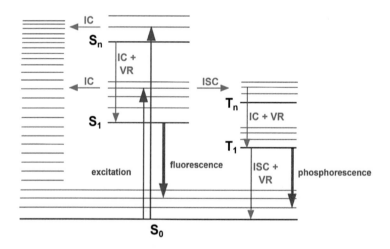

Figure 1.1: Jabłoński-type diagram of possible processes after photoexcitation of molecules in their singlet ground state (S_0) into the first (S_1) or a higher excited singlet state (S_n). Radiationless depopulation processes of the initially excited state can occur via internal conversion (IC) and intersystem crossing (ISC) accompanied by vibrational relaxation (VR). Furthermore, the molecule can be deactivated by fluorescence and phosphorescence.

Typical time constants for IC processes are between 10^{-14}-10^{-4} seconds and for ISC processes between 10^{-12}-10^{-2} seconds indicating that internal conversion is normally faster than intersystem crossing. As illustrated in figure 1.1, all processes described above are accompanied by vibrational relaxation (VR) within the electronic states which occurs on the fs to ps time scale. Additionally, excited molecules can relax via radiative processes. According to Kasha's rule[32], photon emission takes place from the lowest vibrational state of S_1 (fluorescence) and T_1 (phosphorescence). The latter process is on the order of mircoseconds (μs), whereas fluorescence arises on the order of ns. Further processes, which are not included in the Jabłoński diagram, are direct energy transfer from excited states to surrounding molecules as well as chemical reactions involving different excited states.

As has been demonstrated by numerous experimental and theoretical studies[33-39], especially compounds revealing intramolecular charge transfer (ICT) in their initially excited state are characterized by unique and manifold relaxation behaviors. In this regard, molecular structural changes induced by ICT were observed for several donor-acceptor molecules in which the electron donor and the electron acceptor moieties are linked by an aromatic component leading to a planar geometry in the electronic ground state. After intramolecular charge transfer excitation, rotation of the donor moiety out of the aromatic ring plane, so-called twisted ICT, can lead to different equilibrium structures of the ICT state.[35] One of the first clearly recognized and most discussed examples of this class was 4-(N,N-dimethylamino)-benzonitrile (DMABN) that emits two fluorescence bands caused by a 90° twisted conformation.[35, 40] This phenomenon is commonly known as dual fluorescence and can be rationalized noticing that decoupling of the donor and acceptor moieties by out-of-plane rotation leads to minimization of the energy, and thus stabilization of the locally excited ICT state.[33, 41, 42] As a result, two distinguishable fluorescence bands originating from the locally excited and the twisted ICT state are observable depending on the solvent polarity.[35]

Furthermore, ICT has been identified as a trigger for excited state intramolecular proton transfer reactions by an appropriate arrangement of the involved functional groups. As observed in 2-butylamino-6-methyl-4-nitropyridine-N-oxid (2B6N)[36, 43], interaction between the N-oxide and the amino substituent in *ortho* position results in ultrafast proton transfer and generation of the tautomeric form in the excited state. Mediated by intramolecular charge redistribution, this proton transfer occurs on a time scale around 100 fs.[36]

The present work focuses on the ultrafast dynamics after ICT excitation of nitrophenols and their corresponding phenolates (see chapter 4) as well as selected type-I photoinitiators (see chapter 5) by means of femtosecond pump-probe spectroscopy in the liquid phase. For nitrophenols, ICT takes place from an electron donor to an electron acceptor substituent. Hence, nitrophenols serve as excellent model compounds for studying the interaction between donor and acceptor moieties as well as the influence of isomerism and deprotonation facilitated by their small molecular size. On the contrary, in the case of the investigated type-I photoinitiators, ICT occurs from a hydroxybenzyl and morpholino moiety, respectively, to a benzoyl fragment. Such molecules are commonly used as radical precursors since they decompose by α-cleavage of a C–C bond and finally induce polymerization reactions.

2 Fundamentals

This chapter aims at a brief introduction into the fundamentals which are of relevance for the present work and is divided into two subsections: In the first part (section 2.1), optical and experimental basics are discussed including properties, generation, modification as well as application of femtosecond laser pulses. The second part (section 2.2) provides an overview of the principles of the quantum chemical calculations employed in this work.

2.1 Femtosecond time-resolved spectroscopy

2.1.1 Femtosecond pulse properties

Electromagnetic waves are described by the Maxwell equations with respect to their electric and magnetic field.[44] Light pulses in particular can be considered as "wave packets" characterized by an envelope and a carrier frequency as shown in figure 2.1. In most cases, the pulse shape is expressed by a Gaussian, Lorentzian or $sech^2$ function. Hence, a mathematical representation of the electric field (real part) in the time domain is given by equation (1), where E_0 is the amplitude, Γ the shape factor of a Gaussian envelope, ω_0 the carrier frequency and $\Phi(t)$ the temporal phase function.[45]

$$E(t)=\text{Re}\left[E_0 e^{-\left(\Gamma t^2 + i(\omega_0 t - \Phi(t))\right)}\right] \tag{1}$$

The latter contains information about modification of the light pulse propagating through a transparent, dispersive material. An analog representation in the frequency domain is obtained by Fourier transformation.

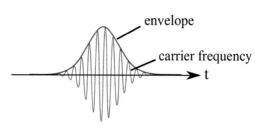

Figure 2.1: Electric field of a Gaussian-shaped laser pulse in the time domain modified according to reference 46.

In order to create femtosecond (fs) laser pulses, an active laser medium with a broad fluorescence spectrum is required. The relationship between the temporal ($\Delta\tau$) and spectral (Δv) full width of half maximum (FWHM) of the pulse is specified by the so-called time-bandwidth product.[45, 47]

$$\Delta(\tau)\,\Delta(v){\geq}K \tag{2}$$

In analogy to the Heisenberg uncertainty-relation, this product can be interpreted as the quantum-mechanical time-energy uncertainty principle, where K is a value depending on the pulse shape as listed in table 2.1. When the equality is reached in equation (2), the pulse is said to be in its Fourier- or transform-limited and $\Delta\tau$ cannot be further reduced.

Traveling through a transparent medium, a laser pulse undergoes a phase distortion due to dispersion phenomena caused by the frequency dependence of the refractive index $n(v)$. The latter connects the velocity of light propagation through a medium ($c(v)$) with the velocity in vacuum (c_0) as shown below.

$$n(v){=}\frac{c_0}{c(v)} \tag{3}$$

As can be derived from Taylor expansion of the frequency dependent phase function, different dispersion effects are distinguished regarding their order.[47] The first-order dispersion, or so-called group delay (GD), has no effect on the pulse shape. In contrast, the second-order dispersion (group delay dispersion, GDD) causes a linear frequency chirp in the pulse. As a consequence, the pulse shape is maintained (e.g. Gaussian-shaped), but the pulse duration $\Delta\tau$ is increased. When the blue components of the pulse are delayed with regard to the red components one speaks of a positive chirp or an up-chirped pulse - as e.g. by traveling through glass ($n_{blue} > n_{red}$).

Taken into account the propagation of two fs pulses having different central wavelengths, the pulses become separated due to their various group velocities. The relative delay of one pulse related to the other is known as group-velocity mismatch (GVM) and has to be considered when interaction of the pulses is required.

Table 2.1: Time-bandwidth product K for selected pulse shapes.[45]

shape	K
Gaussian	0.441
Lorentzian	0.142
sech^2	0.315

2.1.2 Nonlinear optical effects

In the linear regime, the superposition principle can be applied to electromagnetic waves and the optical parameters of matter are independent of the light intensity. In other words, the relationship between the macroscopic polarization of a medium P and the interacting electric field E can be described in a linear term.[44, 47] While developing intense light sources resulting in radiation intensities larger than 10^3 W/cm^2 access to nonlinear optical effects has been facilitated.[47] In this case, the properties of matter strongly depends on the intensity of the incoming light beam and the correlation between P and E has to be expressed in a series expansion as shown in equation (4)[45, 48]

$$P_i(t) = \varepsilon_0 \sum_j \chi_{ij}^{(1)} E_j(t) + \varepsilon_0 \sum_{jk} \chi_{ijk}^{(2)} E_j(t) E_k(t) + \varepsilon_0 \sum_{jkl} \chi_{ijkl}^{(3)} E_j(t) E_k(t) E_l(t) + \ldots$$

$$= P_i^{(1)}(t) + P_i^2(t) + P_i^{(3)}(t) + \ldots$$

$$= P_i^L(t) + P_i^{NL}(t) \quad . \tag{4}$$

Here, ε_0 is the vacuum permittivity and $\chi_{ij}^{(1)}$, $\chi_{ijk}^{(2)}$ and $\chi_{ijkl}^{(3)}$ are tensor components of the electric susceptibility of the medium. Thereby, $\chi^{(1)}$ is a second-order, $\chi^{(2)}$ a third-order and $\chi^{(3)}$ a fourth-order tensor. The first term $P_i^L(t)$ represents the linear optical regime, whereas the second term $P_i^{NL}(t)$ contains all nonlinear responses of the polarization regarding the impinging light. Since nonlinear effects cover a large variety of applications in femtosecond laser spectroscopy, the most prominent and frequently used ones are discussed in the next two sections. In general, all nonlinear processes proceed under energy and momentum conservation.

2.1.2.1 Second-order processes

The second-harmonic generation (SHG) was the first nonlinear optical effect to be experimentally observed by Franken et al. in 1961 focusing a ruby laser pulse onto a quartz crystal.[49] In terms of the photon representation, SHG is described as a mixing process involving three photons. Two photons with the angular frequency ω are up-converted resulting in a single photon having twice the original frequency (2ω). From a more physical point of view, generation of the second-harmonic can be best described by assuming an electric field in a single direction $E_x(t)$ according to equation (5).

$$E_x(t) = E_x e^{-i\omega t} + E_x^* e^{i\omega t} \tag{5}$$

Herein, the field is represented by a harmonic oscillation with a single angular frequency ω and the amplitude E_x, i.e. by a linearly polarized, monochromatic wave. Considering only second-order nonlinearities, equation (4) can be simplified as shown below.

$$P_i^2(t)=\varepsilon_0 \sum_{jk} \chi_{ijk}^{(2)} E_j(t)E_k(t) \qquad (6)$$

By inserting equation (5), the polarization $P_i^2(t)$ then yields

$$P_i^2(t)=\varepsilon_0 \chi_{ixx}^{(2)} \underbrace{\left(E_x^2 e^{-i2\omega t}}_{\textbf{SHG}} + 2E_x E_x^* + c.c.\right) \qquad (7)$$

As one can see, the first term in the bracket oscillates with twice the frequency of the incoming light, and thus is attributable to SHG.[48] The second term is a constant contribution corresponding to a static electric field and commonly known as optical rectification.[44] For the sake of clarity, complex conjugated contributions are summarized and abbreviated with c.c. In the specific case in which the medium is characterized by an inversion center, the second-order susceptibility $\chi_{ixx}^{(2)}$ is zero and no SHG can be observed. Thus, the second-harmonic can only be generated in materials which reveal optical anisotropy, e.g. non-centrosymmetric crystals. A major difficulty in achieving high doubling efficiencies is the aforementioned frequency dependence of the refractive index (section 2.1.1) and therefore the dispersion. Traveling through the medium, the original wave (frequency ω) continuously produces the second-harmonic (2ω). The latter, however, is only macroscopically detectable if previously and newly generated contributions constructively interact; by destructive interference the SHG vanishes. Efficient SHG can be accomplished by a procedure commonly known as phase-matching.[45, 47] Evidently, perfect phase-matching is achieved when the condition $n(\omega) = n(2\omega)$ is fulfilled. In doing so, the most common way to overcome dephasing is to use birefringent crystals. Such crystals are characterized by an optical axis (i.e. by optical anisotropy) and different refractive indices depending on polarization and propagation direction of the fundamental light.[44] When unpolarized light travels through a birefringent crystal, it is split into two components, the so-called ordinary (o) and extraordinary (e) waves.

The latter are defined by a polarization plane parallel to the optical axis, whereas o-waves are polarized perpendicular to the optical axis. Moreover, birefringent crystals have the property that if the incoming light is a linear polarized o-wave, the generated second-harmonic will be a linear polarized e-wave. Figure 2.2 illustrates the index surfaces for a fundamental o-wave having the frequency ω and the resulting e-wave with 2ω. Taking into account the previously mentioned phase-matching condition, efficient SHG is achieved when $n_o(\omega) = n_e(2\omega)$. As can be seen from figure 2.2, this requirement is fulfilled when the index surfaces of the o- and e-waves cross one another characterized by the phase-matching angle θ. Thereby, θ is defined as the angle between the optical axis of the crystal and the propagation direction of the incoming light. Thus, by turning the angle θ, the propagation velocities of the fundamental and second harmonic wave can be matched.

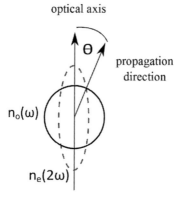

Figure 2.2: Index surfaces for a fundamental o-wave ($n_o(\omega)$, frequency ω, solid line) and the generated e-wave ($n_e(2\omega)$, frequency 2ω, dashed line) in a birefringent crystal modified according to reference 47. The phase-matching condition is fulfilled for the angle θ.

However, considering the aforementioned spectral width of fs pulses (section 2.1.1), the phase-matching condition has to be accomplished for the entire spectral range in order to yield frequency-doubled fs pulses. For this application, the use of thin crystals has turned out to be the best way as the influence of destructive interference is kept to a minimum.[45]

Despite SHG, the second-order susceptibility is responsible for a large variety of further effects implying three photons. Considering an electric field in a single direction $E_x(t)$

consisting of the two different frequencies ω_1 and ω_2, as shown in equation (8), the corresponding expression for the polarization $P_i^2(t)$ is given by equation (9).

$$E_x(t) = E_{1,x}e^{-i\omega_1 t} + E_{2,x}e^{-i\omega_2 t} + \text{c.c.} \tag{8}$$

$$P_i^2(t) = \varepsilon_0 \chi_{ixx}^{(2)} \underbrace{\left(E_{1,x}^2 e^{-i2\omega_1 t}}_{\textbf{SHG } \omega_1} + \underbrace{E_{2,x}^2 e^{-i2\omega_2 t}}_{\textbf{SHG } \omega_2}\right.$$

$$\underbrace{2E_{1,x}E_{2,x}e^{-i(\omega_1+\omega_2)t}}_{\textbf{SFM}} + \underbrace{2E_{2,x}^*E_{1,x}e^{-i(\omega_1-\omega_2)t}}_{\textbf{DFM}} + 2E_{1,x}^*E_{1,x} + 2E_{2,x}^*E_{2,x} + \text{c.c.}) \tag{9}$$

The first and second term in the bracket of equation (9) is associated with SHG of the two fundamental angular frequencies ω_1 and ω_2. The third and fourth term gives rise to photons characterized by the frequencies ($\omega_1 + \omega_2$) and ($\omega_1 - \omega_2$), respectively. In the latter case it is assumed that ω_1 is larger than ω_2. Hence, interaction of two primary waves having different frequencies can further lead to sum frequency mixing (SFM) and difference frequency mixing (DFM).[45, 47] In the picture of photons, two photons with ω_1 and ω_2 are merged during SFM. On the contrary, one photon with ω_1 is annihilated during DFM finally resulting in an additional photon of ω_1 and a photon of the remaining frequency ($\omega_1 - \omega_2$). Since light of the lower frequency is amplified in this process it is also known as optical parametric amplification (OPA). The remaining terms of equation (9) are related to optical rectification and complex conjugated contributions.

Generally, highly nonlinear and birefringent materials such as β-barium borate (BBO, β-BaB$_2$O$_4$), lithium borate (LBO, LiB$_3$O$_5$) and potassium titanyl phosphate (KTP, KTiOPO$_4$) are applied for tunable frequency conversion.[47] In the present work, optical parametric amplification is used for frequency conversion in the visible and NIR as well as SFM is used for frequency conversion in the UV as described in detail in sections 2.1.4 and 3.2.1, respectively.

2.1.2.2   Third-order processes

The number of observable nonlinear effects rapidly grows with the order of the susceptibility.[45, 47] Considering third-order susceptibility, four photons are involved and the polarization of a material can be expressed as displayed in equation (10).

$$P_i^3(t) = \varepsilon_0 \sum_{jkl} \chi_{ijkl}^{(3)} E_j(t) E_k(t)\, E_l(t) \tag{10}$$

In contrast to second-order effects, third-order processes are observable in centrosymmetric materials and no longer restricted to birefringent crystals. They are detectable in liquids and amorphous materials, e.g. fused silica. In analogy to SHG, generation of the third harmonic (3ω, THG) is possible. Since the efficiency of third-order effects, however, is generally very poor they are limited in application. Another way to describe third-order nonlinearities is given by the intensity dependence of the refractive index n according to equation (11)

$$n = n_0 + \frac{1}{2} n_2 I \quad , \tag{11}$$

where n_0 is the linear and n_2 the nonlinear index of refraction and I the intensity of the incoming beam. The intensity dependence of n leads to self-focusing of intense laser pulses which is known as the optical Kerr effect and the basis for self-mode-locking as occurring in Ti:Sa lasers.[47] Moreover, considering the time dependence of the pulse intensity, spectral broadening is caused due to self-phase-modulation (SPM). Thereby, new low frequencies are generated in the leading edge and new high frequencies are created in the trailing edge of the pulse envelope (assuming that n_2 is positive).[47] One common used application of SPM is the generation of white-light continua (section 2.1.5) from fs laser pulses which allows for tunable frequency conversion (section 2.1.4) or transient absorption spectra (section 3.3).

2.1.3 Application, generation and modification of femtosecond laser pulses

2.1.3.1 Femtosecond pump-probe spectroscopy

The study of photophysical and -chemical phenomena has attracted a lot of interest in many fields like physics, chemistry and biology.[1] Since excited state evolution occurs on an ultrafast time scale (see indroduction), time-resolved techniques have become a powerful experimental approach, e.g. fs pump-probe absorption spectroscopy which was employed in the present work. The general principles of this technique are explained in the following. A fs light pulse, the so-called pump pulse, perturbs the sample by excitation into higher electronic states as demonstrated on the left part of figure 2.3. Subsequently, a second fs pulse, the so-called probe pulse, crosses the perturbed sample and acts as a probe for the generated excited species. In order to monitor the excited state evolution with respect to the time scale, the probe pulse is delayed relative to the pump beam. The probe pulse itself can consist of a single central wavelength or can cover a wide spectral region (white light continuum, see section 2.1.5). To detect only the spectral signature of the excited species, and not of molecules in the electronic ground state, the transmitted probe light through the pumped (excited) and unpumped (unexcited) sample is measured. Thus, the change of the optical density (ΔOD), also referred to as transient absorbance (TA), is given by

$$\Delta OD(t) = OD(t)_w - OD_{\frac{w}{0}} \ , \tag{12}$$

where $OD(t)_w$ is the optical density with and $OD_{\frac{w}{0}}$ without previous excitation through the pump pulse. Thereby, $\Delta OD(t)$ is obtained by consecutive probe pulses and $OD_{\frac{w}{0}}$ is time-independent. When the generated excited species absorb at the probe wavelength, the change of OD becomes positive, commonly known as excited state absorption (ESA). On the contrary, the excited species can relax to the ground state under the emission of photons. This process is induced by the probe pulse and therefore referred to as stimulated emission (SE). Another contribution to ΔOD having the same sign is given by ground state bleaching which can be observed if the probe and pump wavelengths are similar to each other. In this case, depopulation of the ground state by the pump pulse leads to a reduction of an additional excitation by the probe pulse. Furthermore, already relaxed but vibrational highly excited ("vibrational hot") molecules in the electronically ground state can absorb at the probe wavelength, and thus lead to positive ΔOD values. Since the absorption spectrum of such molecules is red-shifted compared to the ground state spectrum, contributions of hot ground state absorption (HGSA) are normally detected red-shifted with respect to the ground state absorption. The discussed processes are summarized in figure 2.3 on the right part. In general, transient absorption consists of a convolution of the described contributions, especially considering the similar time scale of many relaxation processes (see introduction).

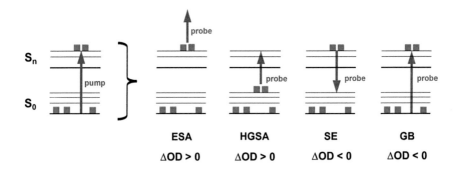

Figure 2.3: Generation of excited molecules due to excitation by a pump pulse (left side). The molecules are represented by grey squares. Probing leads to different contributions finally resulting in positive or negative ΔOD values. These processes are summarized on the right side, namely excited state absorption (ESA), hot ground state absorption (HGSA), stimulated emission (SE) and ground state bleaching (GB).

2.1.3.2 Generation of femtosecond laser pulses

The laser system (CPA 2210, Clark-MXR) employed in the present work is based on passive mode locking.[10] The generated pulses are specified by a temporal width between 150 and 170 fs and an energy of 1.6 mJ/pulse, as the laser power is 1.6 W with a repetition rate of 1 kHz.[50] In the following, the four main components of the system are briefly illustrated, namely ring fiber oscillator, pulse stretcher, regenerative amplifier and compressor.[51]

The basic design of the ring fiber oscillator has been developed by the group of Ippen and Haus *et al.* using an Er^{3+}-doped fiber as gain medium with an emission maximum at

1550 nm.[52] The fiber itself is pumped by a laser diode operating at approximately 980 nm. Exploiting the nonlinearity of the refraction index (section 2.1.2.2), laser pulses are produced by additive pulse mode locking which is self-adjusting.[53, 54] Thereafter, the second-harmonic of the pulses is created using a KTP crystal in order to achieve a central laser wavelength of 775 nm. However, since the pulse energy is in the order of only a few nanojoules, it has to be increased. In doing so, a regenerative amplifier[55] is used consisting of a Ti:Sa crystal that is pumped by a Q-switched, frequency-doubled Nd:YAG laser at 532 nm operated at ~10 W. In order to avoid damage of optical components and the occurrence of undesirable and uncontrollable nonlinear effects due to too high pulse intensities the laser pulses are stretched with respect to their duration before injection into the amplifier. This method is known as chirped pulse amplification (CPA) and allows for enhanced amplification efficiencies.[56, 57] The schematic setup of the regenerative amplifier is shown in figure 2.4. For injection of the seed pulse (oscillator pulse) two Glan-Taylor polarizers are employed which reflects the pulse dependent on its polarization plane. The first polarizer (P1) directs the seed pulse through a Faraday isolator; the second polarizer (P2) couples the pulse into the amplifier. The Faraday rotator itself changes the polarization plane of the incoming light dependent on its traverse direction. Within the resonator the seed pulse circulates until gain saturation is reached, controlled by a Pockels cell. The amplified pulses are coupled out again reflected by the Glan-Taylor polarizers. Finally, the pulses pass through a transmission grating which operates as pulse compressor, and thus compensates for the previous chirp and additional dispersion arising during amplification.

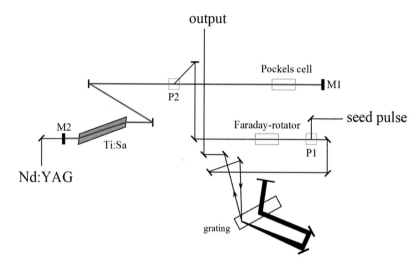

Figure 2.4: Scheme of the regenerative amplifier and compressor. The beam path in the resonator is marked by red lines. M1 and M2 are the end mirrors of the resonator. P1 and P2 are Glan-Taylor polarizers. For details, see text above.

2.1.4 Frequency conversion in the visible and near infrared region

Frequency conversion in the visible and near infrared region is performed by OPA (see section 2.1.2.1) using a commercial non-collinear phase-matched optical parametric amplifier (NOPA, Clark-MXR). Therefore, a portion of the 775 nm CPA 2210 output (~250 µJ/pulse) is split into two parts by a 1% beam splitter. Subsequently, the smaller part is focused into a sapphire plate (3 mm) in order to generate a white light (WL) continuum which covers the spectral range from 470-700 nm in the visible and from 900-1600 nm in the NIR. Since the area between 750-870 nm is featured by an irregular spectral structure due to the degeneracy with respect to the central laser wavelength, it cannot be used for frequency conversion.[14, 15, 58] The remaining part of the 775 nm beam is frequency-doubled in a BBO crystal and split by a 50% beam splitter. In order to amplify certain wavelengths (signal photons) the WL is temporally and spatially overlapped with one portion of the second harmonic (pump photons) in a BBO crystal (1 mm). The amplification process itself is based on second-order nonlinearities as described in section 2.1.2.1. One pump photon is annihilated while generating one signal photon and one photon having the remaining frequency ($\omega_{pump} - \omega_{signal}$).

By changing the temporal overlap as well as the tilt of the BBO crystal with respect to the incoming light, tunability of the signal light can be obtained. Furthermore, the non-collinear arrangement of the pump and signal beams allow for compensation of the group velocity mismatch (section 2.1.1), and thus for an increase of the amplification efficiency. In order to achieve energies up to 7 µJ in the visible, and a few µJ in the NIR, a second amplification process is required. In analogy to the previously described procedure, the already existing signal light is focused into a second BBO crystal (2 mm) along with the remaining part of the second harmonic. Since for many molecular systems the significant transitions are in the UV, frequency conversion within this region is of particular interest. Techniques for UV pulse generation as employed in the present work are described in detail in section 3.2.1 and 3.3.

2.1.5 Generation of white light continua

In general, white light (WL) continua are generated by focusing ultrashort laser pulses under proper conditions into a nonlinear medium, as e.g. liquids, photonic crystal fibers and solids.[59] The corresponding features of the continua, however, strongly depend on the way they are produced. Especially for time-resolved studies, the use of continua obtained from bulk materials like CaF_2 and sapphire have become standard for probing the samples.[60-64] Such continua are characterized by a broad and smooth spectrum, a high temporal and spatial coherence as well as a high pulse-to-pulse stability.[59] Since WL generation is based on self-phase-modulation[13] and therefore corresponds to third-order nonlinear processes (see section 2.1.2.2), the seed pulse has to be strongly focused into the crystal.[65, 66] As displayed in figure 2.5, a continuum obtained from CaF_2 allows for a higher amount of UV components in comparison to a white light generated in sapphire at a seed wavelength of 775 nm. The useful spectral range extends from 420-720 nm for sapphire and from 290-720 nm for CaF_2.[59] The latter is characterized by an almost flat plateau, and thus by similar intensities over a large spectral area. On the contrary, the continuum obtained in sapphire shows an increase in intensity around 500 nm and a decrease for red components. Both spectra are featured by spectral modulations near the fundamental indicated by the sharp intense peak around 720 nm.

Employing probe pulses that cover a wide spectral region at once can greatly improve the understanding of photophysical and -chemical relaxation processes. Since more spectral signatures of the transient species can be monitored simultaneously, a more detailed insight into the underlying relaxation mechanism is provided. Nevertheless, a white light continuum cannot be compressed to the Fourier limit, e.g. NOPA-generated pulses. As a consequence, the experimental time resolution is mostly larger for probing with a WL continuum in comparison to selected probe wavelengths (section 3.5.1). As demonstrated in reference 67, a sapphire-generated WL also contains NIR components. The intensities of these contributions, however, are rather weak, if a seed pulse with a pulse duration of ~150 fs is employed. Nonetheless, one can generate more intensive continua in the NIR with such laser systems by increasing the seed wavelength.[68, 69]

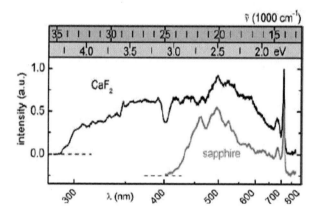

Figure 2.5: White light continua generated in CaF_2 and in sapphire taken from reference 59. The dip around 400 nm in the CaF_2 WL is caused by the used dielectric mirror to suppress the fundamental.

2.2 Theoretical methods

For evaluation of fs transient absorption data, and thus understanding of the underlying relaxation behavior, it is essential to have information about (excited) state properties as provided by quantum chemical calculations.

In quantum mechanics, atomic and molecular systems in their ground state are described by the time-independent Schrödinger equation

$$\hat{H}\,\psi(\vec{r},\vec{R})=E\,\psi(\vec{r},\vec{R})\quad,\tag{13}$$

where $\hat{H}$ is the Hamiltonian, E the energy of the system and $\psi(\vec{r},\vec{R})$ the wave function depending on the coordinates of the electrons $(\vec{r})$ and nuclei $(\vec{R})$. In most cases equation (13) is solved with the help of the Born-Oppenheimer approximation which treats $\vec{R}$ as a parameter. Thus, separation of the electronic and the nuclear wave function is possible. Considering time-dependent systems, e.g. excited states, the time dependence of the Schrödinger equation has to be taken into account.[70, 71]

In general, two types of calculations can be distinguished: single- and multi-reference methods. Whereas single-reference methods are only based on the configuration of the electronic ground state, the multi-reference approach involves a set of different electronic configurations. In the following sections, the two theoretical methods employed in the present work, (time-dependent)-density functional theory (single-reference) and complete active space self-consistent field methodology (multi-reference) are briefly introduced.

2.2.1 (Time-dependent)-Density functional theory

One of the fundamental postulates of quantum mechanics asserts that microscopic systems can be completely described by a wave function that contains all information about the physical properties of this system. The wave function itself, however, does not represent any observable and should be considered as a purely mathematical expression. To avoid the construction of such a function the basic idea of the density functional theory is to form the Hamiltonian with a useful physical observable, namely the electron density ρ as proposed by Hohenberg and Kohn.[71] Integration of ρ over the entire space (spatial coordinate $\vec{r}$) yields the total number of electrons N (equation (14)) and the assignment of the nuclear atomic numbers of the system is the only issue in order to fully characterize the Hamiltonian.

$$N=\int \varrho(\vec{r})d\vec{r}\tag{14}$$

Hence, the energy of the system can be obtained as a functional of the electron density. Early approximations to determine the molecular energy, however, raised the problem of how to create an appropriate expression for ρ. A turning point in the use of DFT was marked by Hohenberg and Kohn in 1964 in terms of the proof of two theorems which are known as the Hohenberg-Kohn existence theorem and the Hohenberg-Kohn variational theorem.[72] Moreover, in 1965, Kohn and Sham derived an expression for the electron density assuming a non-interacting system of electrons which is specified by its ground state Slater determinant.[73] In this so-called Kohn-Sham self-consistent field methodology, the Kohn-Sham one-electron operator is defined as

$$\hat{h}_i = -\frac{1}{2}\nabla_i^2 + \sum_k^{nuclei} \frac{Z_k}{|r_i\text{-}r_k|} + \int \frac{\varrho(r^{'})}{|r_i\text{-}r^{'}|} dr^{'} + V_{xc} \quad , \tag{15}$$

where Z is the atomic number and k run over the nuclei. The terms on the right hand side of equation (15) refer to the kinetic energy of the non-interacting electrons, the nuclear electron interaction, the classical electron-electron repulsion and the so-called exchange-correlation term V_{xc}. The latter contains the correction to the kinetic energy regarding the interacting nature of the electrons as well as all non-classical corrections to the electron-electron repulsion energy. Over the years, various methods have been developed in order to derive expressions for the exchange-correlation term, e.g. the local density and generalized gradient approximation.[74, 75] One of the most widely used functionals today is B3LYP (Becke, three-parameter, Lee-Yang-Parr)[76, 77] a so-called hybrid functional.

Properties of excited molecules, as generated in the presence of a time-dependent potential such as an electric field, can be computed using time-dependent DFT (TD-DFT).[71] The formalism behind TD-DFT is known as Runge-Gross theorem, which is the time-dependent analog of the Hohenberg-Kohn theorem.[78, 79] As demonstrated in several theoretical studies, TD-DFT allows for accurate computation of excitation energies with respect to states characterized by singly excited character. Thereby, a low computational effort facilitates even calculation of large molecules whose size precludes the use of large multi-reference methods, like the photoinitiators investigated in the present work (see chapter 5). However, since TD-DFT belongs to the single-reference methodologies, it is limited in describing excited states in the vicinity of state crossings, as in the case of conical intersections.[80] Furthermore excited states revealing doubly excited character are poorly handled.[80]

2.2.2   Complete active space self-consistent field method

The complete active space self-consistent field method (CASSCF) belongs to the post-Hartree-Fock (post-HF) procedures, and is thus based on the foundations of the Hartree-Fock self-consistent field methodology (HFSCF) and in particular on the configuration interaction method.[70, 71] The latter employs a set of Slater determinants as wave function which consists of the determinant of the electronic ground state as well as of determinants corresponding to excitations of the ground state configuration. Thereby, single, double, triple and so on excitations are distinguished and one speaks of a full configuration interaction if all possible excited configurations are considered. Moreover, this set of Slater determinants can be expanded to excitations related to some excited state configurations and is therefore known as multi-reference configuration interaction. However, since even for small molecules full configuration interaction calculations require a high expenditure of time as well as a high computational demand truncated approaches are commonly chosen. Reducing the number of determinants, it is important how to handle the excitations included in the calculation or in other words how to go about selecting the orbitals that should be allowed to be partially occupied. A typical approach is to define a complete active space (CAS) which contains a limited number of electrons (m) and orbitals (n), often abbreviated as (m, n)-CAS. In this space, all possible occupation schemes are allowed. Over time, various procedures have been developed in order to speed up CASSCF calculations.[71] As illustrated in figure 2.6, one possibility is to minimize the size of the CAS, but to allow a selected number of excitations that involve orbitals outside of this CAS in return. This secondary space is called restricted active space (RAS) and the excitation level is usually limited to single or double excitations. Furthermore, the remaining occupied (normally core orbitals) and unoccupied orbitals are restricted to double or zero occupation, and thus reckoned as frozen. Typically, a certain amount of trial and error is required in order to create an appropriate and stable active space. In opposite to TD-DFT, however, CASSCF calculations allow for describing the potential energy surfaces (PESs) even in the vicinity of state crossings due to their multi-reference character. In these cases, especially so-called state-averaged calculations yield reasonable results.[70] Since relaxation via conical intersection is of major importance for *ortho*-nitrophenol (see section 4.2) as well as the nitrophenolates (see section 4.4), calculations at the CASSCF level of theory were performed in the present work.

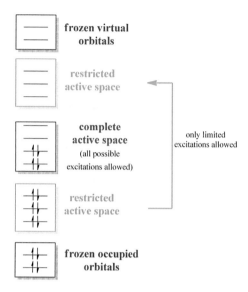

Figure 2.6: Assignment of different spaces in a general multiconfiguration mechanism modified according to reference 70. The complete active space is defined as (4, 4)-CAS consisting of 4 electrons and 4 orbitals.

3 Experimental conditions

3.1 Steady state spectroscopy

Steady state UV-Vis absorption spectra in solution were performed in fused-silica cuvettes (Hellma Analytics) with an optical path length of 1 mm by means of a Cary 500 UV-Vis-NIR spectrometer (Varian). The spectra were normalized to the concentration of the samples in accordance with the Beer-Lambert law[81] as shown in equation (16)

$$\varepsilon(v) \left[\frac{L}{mol \cdot cm} \right] = \frac{OD}{c \cdot d} \quad , \tag{16}$$

where $\varepsilon(v)$ is the frequency dependent molar decadic extinction coefficient, OD the optical density, c the concentration of the sample in mol/L, and d the optical thickness of the cuvette in cm.

3.2 Transient absorption spectroscopy using selected probe wavelengths

3.2.1 Frequency conversion in the ultraviolet

Femtosecond pump pulses in the UV are generated by SFM using a self-built UV-NOPA system (according to reference 82) which is shown in figure 3.1. To obtain tunable pulses between 300 and 370 nm a commercial NOPA system (Clark MXR) is operated in the visible region pumped by 250 µJ of the CPA 2210 output. These NOPA pulses achieve energies of 5-10 µJ/pulse dependent on wavelength and are sent to a BBO crystal (100 µm, $\theta = 35°$,

Döhrer) by a concave aluminum mirror (f = 400 mm, Laser Components) without previous compression (pulse duration >200 fs). Additionally, another portion of the 775 nm laser output (100 μJ, ~150 fs) is directed to the BBO crystal by a plan-convex lens (f = 300 mm, 775 nm anti-reflex coating, Laser Components). To obtain a high efficiency of the SFM process, the foci of both beams are nearly collinearly overlapped using a silver half-mirror in front of the BBO crystal. The crystal itself is placed about 12 mm before the joint focus of both beams in order to avoid damaging as well as higher-order effects induced by too high peak intensities. Considering the difference of the temporal pulse width of the uncompressed NOPA and the 775 nm beam, a higher SFM efficiency is also obtained by enhancing the temporal overlap. In doing so, the 775 nm pulse is directed through a glass block (SF10, 45 mm) and thereby chirped before sending to the BBO crystal. A manual delay stage is employed to control the temporal overlap of the NOPA and the 775 nm pulses. In this way, UV pulse energies between 0.6 to 1 μJ/pulse are achieved. Furthermore, pulses >400 nm can be generated when operating the NOPA system in the NIR. However, as a consequence of less NOPA output energy and a higher beam diameter in the NIR, the efficiency of the SFM process is significantly reduced, e.g. resulting in pulse energies of 0.3 μJ/pulse at 418 nm. Since the SFM process is only nearly collinear, the generated pulses can easily be separated from the remaining fundamentals and the second harmonic of the NOPA pulses, respectively. Based on the p-polarization of the NOPA and the 775 nm beam the produced pulses exhibit s-polarization. For optimizing, a prism compressor consisting of two fused silica prisms with a special anti-reflex coating for s-polarized pulses and an apex angle of 43° is used. The compressor was adjusted by intensity autocorrelation experiments.[83] Since the commercial autocorrelator (NOPA P.I 009, Clark-MXR) does not allow for determination of pulse durations in the UV, the latter were estimated as described in detail in section 3.4. Afterwards, the pulses are coupled out of the compressor by an aluminum half mirror and sent into the sample.

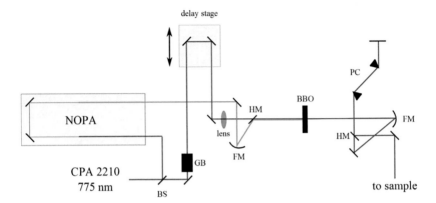

Figure 3.1: Scheme of UV pulse generation using SFM with BS (beam splitter), GB (glass block), FM (focusing mirror), HM (half-mirror), BBO (β-barium borate crystal) and PC (prism compressor). For a detailed description, see text above.

3.2.2 Experimental setup

The employed experimental setup is sketched in figure 3.2. As described before (section 3.2.1), the pump pulses are generated by frequency conversion in the UV and subsequently focused into the sample by a concave aluminum mirror (f = 500 mm, Laser Components). For probing the sample, a second NOPA system (Clark-MXR) is operated in the visible and NIR region pumped with 250 µJ of the CPA 2210 output. To compensate for the group delay dispersion of the probe pulses, mainly caused by the NOPA process itself, a prism compressor is used. In doing so, adjustment of the compressor was performed by intensity autocorrelation experiments at each probe wavelength. The FWHM of the optimized probe pulses was determined to be at 27-35 fs dependent on the wavelength. After passing through a delay line, the probe pulses are directed to the sample by a concave aluminum mirror (f = 400 mm, Laser Components) and spatially overlapped with the pump beam. The delay line consists of a computer controlled translation stage (Physical Instruments) with a maximum length of 30 cm corresponding to 2 ns. This controls the time shift between the pump and probe pulses. Two photodiodes (Hamamatsu), labeled as PD1 and PD2 in figure 3.2, are placed in front of and behind the sample cell which allow for measuring the intensity I_0 and I of the probe beam, and thus for detection of the OD ($\log \frac{I_0}{I}$) of the sample. Furthermore, a chopper wheel (Ametek) is inserted into the beam path of the pump pulses and synchronized with half of the laser frequency (500 Hz) to reduce the repetition rate. A third photodiode (PD3, Hamamatsu) behind this wheel records if the pulses are transmitted or not. In this way, the time-dependent change in OD is yielded as described below

$$\Delta OD(t) = OD(t)_w - OD_{\underset{o}{w}} = \log\left(\frac{I_0}{I(t)}\right)_w - \log\left(\frac{I_0}{I}\right)_{\underset{o}{w}} ,\qquad (17)$$

where $OD(t)_w$ and $OD_{\underset{o}{w}}$ are the optical density with and without previous excitation by the pump pulse, respectively. To avoid the influence of anisotropy all transient absorption (TA) experiments were performed under magic angle conditions. Thereby, the polarization plane of the probe pulses is set to 54.7° with respect to the pump pulse polarization using a λ/2 plate (Spindler and Hoyer). Data acquisition was performed using a home-made software in a Labview environment (National Instruments). The temporal overlap between the pump and probe beam was adjusted by cross correlation experiments in dye solutions (see section 3.5.1).

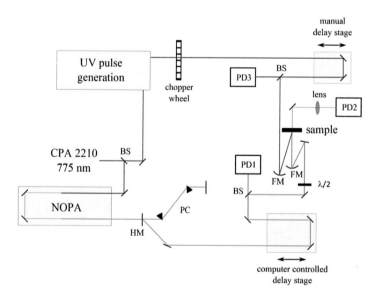

Figure 3.2: Scheme of the experimental setup using selected probe wavelengths with BS (beam splitter), HM (half-mirror), PC (prism compressor), FM (focusing mirror) and PD (photodiode). The path of the pump beam is indicated by blue lines, whereas the path of the probe beam is marked by green lines. For details, see text above.

3.3 Transient absorption spectroscopy using white light continuum

Figure 3.3 shows a scheme of the used experimental setup in accordance with a state-of-the-art fs transient spectrometer described by Riedle *et al.*[59] A portion of the CPA 2210 output (250 µJ) is split into two parts by a 5% beam splitter. The major part is employed to operate a commercial NOPA system (Clark-MXR) in the visible region, whereas the smaller part is used for generation of a WL continuum. The NOPA pulses (4-5 µJ/pulse dependent on wavelength) are subsequently up-converted into the UV directing the beam into a BBO crystal (0.2 mm, $\theta = 35°$, Castech) by a concave aluminum mirror (f = 300 mm, Laser Components). That way, tunable pump pulses between 240 and 350 nm are accessible. In order to achieve a high efficiency of the second order process, and thus high UV pulse energies, the NOPA pulses are dispersion optimized by a prism compressor before frequency doubling.

As has been already mentioned in the previous section (see section 3.2.2), the alignment of the compressor was performed by optimizing the FWHM of the NOPA pulses. As a result, pulse durations between 30 and 40 fs were achieved which correspond to UV pulse energies of 0.25-0.35 µJ/pulse dependent on wavelength. Before passing through a computer controlled translation stage (27 cm corresponding to 1.8 ns, Thorlabs) the second harmonic is separated from the remaining fundamental NOPA light by employing three dielectric mirrors labeled as DM in figure 3.3 (high reflecting for 300-360 nm and high transmitting for 470-750 nm, Laser Components). The delay line itself consists of an aluminum retro reflector (PLX) which allows for parallel adjustment of the beam path with an extremely high accuracy. Then, the pump beam is set to magic angle conditions using a $\lambda/2$ plate (Spindler and Hoyer) and directed to the sample by a concave aluminum mirror (f = 1000 mm, Laser Components).

Even though a broader spectral region is covered by a WL continuum generated in CaF_2 in comparison to that obtained from sapphire (see section 2.1.5), the latter was employed for probing the samples. Since the pulse duration of the laser fundamental increases towards the transient spectrometer (>200 fs) it was not possible to achieve a stable continuum in CaF_2. Therefore, 5% of the original 250 µJ of the laser fundamental are focused into a sapphire plate (1 mm) by a lens (f = 100 mm, Laser Components), labeled with 4 and 3 in figure 3.3, respectively. To produce a stable single filament the intensity of the incoming 775 nm beam was varied by an iris (1) and a continuously variable neutral density filter (2, Edmund Optics). The spectroscopically useful range of the generated WL continuum spans from about 435 to 710 nm. After recollimation, the seed wavelength is removed from the WL by a custom-made dielectric filter (0°, high reflecting for 750-850 nm and high transmitting for 290-690 nm, Laser Components). Subsequently, the probe beam is focused into the sample by a concave aluminum mirror (f = 300 mm, Laser Components) and split in a prism-based polychromator after transmission. The different spectral ranges are detected using a charge-coupled device (CCD) camera (Linescan 2000, Ingenieurbüro Stresing).

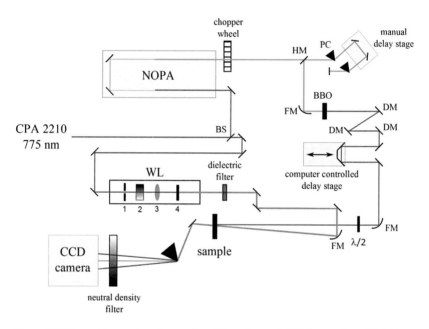

Figure 3.3: Scheme of the experimental setup using a WL continuum with BS (beam splitter), HM (half-mirror), PC (prism compressor), FM (focusing mirror), BBO (β-barium borate crystal) and DM (dielectric mirror). The path of the pump beam is indicated by blue lines, whereas the path of the WL is marked by grey lines. For detailed description of WL generation, see text above.

Thereby, the CCD camera was aligned by interference filters with bandwidths of 50 nm (Thorlabs) which were inserted into the WL beam path at the position of the sample. Finally, the camera pixel coordinates were assigned to the corresponding wavelengths using the Sellmeier equation[59] as implemented in the employed home-made software in a Labview environment. For efficient exploitation of the detector`s dynamical range, the transmitted and dispersed WL is attenuated using a continuously variable neutral density filter in front of the

CCD camera. Figure 3.4 shows a spectrum of the dispersed sapphire WL. In accordance with the literature[59], the spectrum is characterized by a lack of intensity between 600 and 650 nm as well as an increased contribution at 700 nm. The latter leads to intensities close to the saturation level of the CCD camera despite the suppression of the fundamental and the use of the neutral density filter. Spectral fluctuations around 700 nm are caused by the proximity to the seed wavelength.

Furthermore, a chopper wheel (Thorlabs) is placed into the beam path of the pump pulses to reduce the repetition rate to 500 Hz and to allow for measuring the change of the optical density. Since the intensity of the WL is only measured after transmitting through the sample, equation (17) is simplified as shown in equation (18)

$$\Delta OD(t)=OD(t)_w - OD_{\underset{o}{w}} = -\log\left(\frac{I(t)_w}{I_{\underset{o}{w}}}\right),\qquad(18)$$

where $I(t)_w$ and $I_{\underset{o}{w}}$ are the probe light with and without previous excitation by the pump pulse. The temporal overlap between the pump and probe beams was adjusted by cross correlation experiments in dye solutions (see section 3.5 for details). In addition, the recorded data were corrected for the chirp of the WL caused by wavelength dependent group velocity mismatch when passing through the optical components in the probe beam. Thereby, chirp correction, and thus determination of time-zero, was done using the coherent artifact characterized by the temporal overlap of the pump pulse with a certain frequency component of the WL continuum.[59, 84-87]

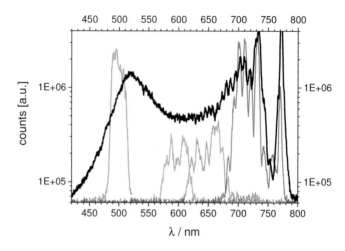

Figure 3.4: Dispersed white light generated in a sapphire plate (black line) and recorded with a CCD camera. The colored lines display different parts of the WL filtered by interference filters with bandwidths of 50 nm (Thorlabs). The high peak around 775 nm is caused by the seed wavelength. Note that the values are plotted on a logarithmic ordinate scale.

3.4 Analysis of transient absorption data

The diameter of the spot size of pump and probe beam were of 2 and 1 mm, respectively, for both experimental setups described in section 3.2.2 and section 3.3. Hence, the sample volume irradiated by the pump pulse was larger than that irradiated by the probe pulse finally leading to a complete exploitation of the probe pulse. Determination of time-zero was performed by cross correlation of pump and probe beams in suitable fluorescence dye solutions with, if possible, the same OD as the samples. Since dye molecules are characterized by long-lived excited states (lifetimes in the order of ns)[88], no significant dynamics are expected on the time scale of the TA experiments. Thus, ΔOD stays constant after time-zero and can be analyzed according to equation (19)

$$\Delta OD(t) = g(t) \cdot C \ , \tag{19}$$

where g(t) is the experimentally determined Gaussian cross-correlation function, which models the increase of the TA profiles, and C is a constant term. Assuming a Gaussian temporal shape of the employed fs pulses, g(t) can be reckoned as a convolution of a step and Gaussian function resulting in an error function as shown in equation (20).[16]

$$g(t)=\frac{1}{2} \cdot \left[1 + erf\left(\sqrt{4 \cdot \ln 2}\right)\frac{t}{\tau_0}\right]$$

(20)

Thereby, τ_0 can be considered as an appropriate measure of the experimental time resolution and is related to the temporal FWHM of the employed pump (Δt_{pump}) and probe (Δt_{pump}) pulses as displayed in equation (21)

$$\tau_0 = \sqrt{\Delta t_{pump}^2 + \Delta t_{probe}^2} \quad .$$

(21)

Hence, the determination of τ_0 and Δt_{probe} (see section 3.5 and 3.2.2) allows for the estimation of the temporal length of the UV pulses to be about 100 fs. In order to analyze the excited state evolution of molecules revealing dynamics on the time scale of the TA experiments, amplitude C in equation (19) has to be substituted by a sum of exponentials. As depicted in equation (22), this sum contains the time constants τ_i of the excited state dynamics as well as the corresponding amplitudes A_i (λ_{probe}) as parameters.

$$\Delta OD\left(t, \lambda_{probe}\right) = g(t) \cdot \left[\sum_{i=1}^{n} A_i\left(\lambda_{probe}\right) \cdot e^{-\frac{t}{\tau_i}}\right]$$

(22)

The latter can be regarded as the spectral signature of different relaxation steps. The amplitudes depend on the probe wavelengths and include information on excited species, as e.g. absorption cross sections. Extraction of A_i (λ_{probe}) from TA profiles probed at various wavelengths is commonly known as so-called decay associated difference spectra (DADS). Positive values are associated with a decrease of ΔOD, whereas negative values are related to an increase of ΔOD. Nevertheless, in order to compare the contributions of specific relaxation channels at different probe wavelengths, relative amplitudes $A_{i,rel}$ (λ_{probe}) have to be determined. In other words, A_i (λ_{probe}) has to be normalized at each probe wavelength as shown in equation (23)

$$A_{i,\,rel}\left(\lambda_{probe}\right)=\frac{A_i\left(\lambda_{probe}\right)}{\sum_{i=1}^{n} A_i\left(\lambda_{probe}\right)} \quad . \tag{23}$$

TA spectra probed with both selected wavelengths and the WL continuum can be analyzed in the same way. A direct comparison between the employed TA setups is exemplarily discussed in section 4.3.2 for the excited state dynamics of *meta-* and *para-*nitrophenol. The following part provides an overview of the experimental differences of both setups. Since more pump energy is generated by the SFM compared to the frequency doubling process (see section 3.2.1 and 3.3), ΔOD is 2-3 times higher for probing with selected wavelengths which leads, in many cases, to a better signal-to-noise ratio. Furthermore, this ratio is also influenced by limited temporal and spectral stability of the generated sapphire WL which cannot be corrected. To avoid decomposition of the sample caused by laser irradiation it is important to keep the total measurement time at a minimum. Despite continuously pumping of the sample, one experiment should not exceed 45-60 minutes as evaluation of steady state spectra before and after the time-resolved measurement clearly indicated. Due to different step size control for data acquisition, the maximum usable delay time was about 1800 ps and 500 ps for the transient and the single wavelength setup, respectively. Implementation of exponential steps in the data acquisition process allows for a longer delay time for WL generated spectra, but at the cost of the experimental step size.

3.5 Samples

All time-resolved experiments were performed in solution at 20°C in a flow cell system consisting of a fused silica cuvette (Hellma Analytics) with an optical path length of 1 mm and a gear pump. Chloroform ($\geq$99.9%), 2-propanol ($\geq$99.9%) and n-hexane ($\geq$98%) were purchased from Carl Roth. Methylisobutyrate (MIB, $\geq$99%) was purchased from Sigma

Aldrich. Deionized water was used from our local source. All solvents were employed without further purification.

3.5.1 Nitrophenols and corresponding phenolates

ortho-, *meta-* and *para-*nitrophenol (*o*-NP, ≥99.5%; *m*-NP ≥99%; *p*-NP ≥99.5%) were purchased from Sigma Aldrich and used without further purification. The samples were prepared as solutions in chloroform, 2-propanol and water. *n*-hexane could only be used for *o*-NP due to low solubility of *m*- and *p*-NP in non-polar solvents. The employed pump wavelengths (λ_{pump}) were 350 nm for *o*-NP, 330 nm for *m*-NP and 314 nm for *p*-NP. The optical density of the samples was set to a value of 3 at λ_{pump} in order to minimize the group velocity mismatch of the pump and probe beam, and thus to improve the experimental time resolution. Therefore, the concentrations were in the range of $(0.1\text{-}1.7)\cdot10^{-2}$ mol/L. TA experiments using selected probe wavelengths (λ_{probe}) were conducted for *o*-NP in a spectral range of 480-1100 nm and for *m*- and *p*-NP between 480 and 950 nm. That way, the smallest steps are chosen to be 10 nm due to limitation of the spectral pulse widths of the probe pulse. Using a transient spectrometer (section 3.3), a more detailed investigation was facilitated for *m*- and *p*-NP. On the contrary, low ΔOD values do not allow for transient spectra of *o*-NP, in particular considering the pump energy generated by frequency doubling compared to the SFM process (see section 3.2.1 and 3.3). The comparison of the data sets recorded for *m*- and *p*-NP with both experimental setups was also used for validation of the transient spectrometer. In doing so, analysis of the data indicates an experimental artifact when probing with the sapphire WL continuum. The transient spectra are featured by a lack of TA around 600 nm and an additional increase of TA around 700 nm, as exemplarily displayed in figure 3.5 for *p*-NP in water. The entire spectral range, however, is characterized by the same dynamics and therefore the observed effect is solely caused by the intensity distribution of the dispersed WL as depicted in section 3.3. As a consequence, WL generated TA data are exclusively shown from 435 to 600 nm in the following. Since TA data recorded with selected probe wavelengths additionally reveal only low TA at λ_{probe} higher than 600 nm, the relevant spectral area is covered with the limited WL.

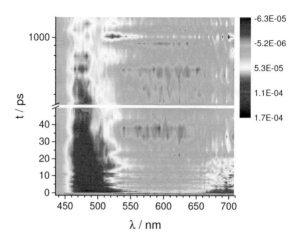

Figure 3.5: TA spectrum of *para*-nitrophenol probed with the sapphire WL. The spectrum indicates a lack of TA around 600 nm. For more details, see text above.

Furthermore, time-resolved experiments of the deprotonated nitrophenols were conducted at 418 nm (λ_{pump}) and at 500-950 nm (λ_{probe}). Thereby, deprotonation was performed in water and chloroform using sodium hydroxide (NaOH) and 1,8-diazabicyclo[5.4.0]undec-7-ene (DBU) as base. DBU was chosen as organic base because it shows no electronic transitions in the visible region, and thus no dynamics at λ_{pump}. NaOH ($\geq$99%) and DBU ($\geq$99%) were purchased from Roth and Sigma Aldrich, respectively. As derived from concentration dependent steady state spectroscopy (appendix A, figure A), complete deprotonation is achieved with a ratio (nitrophenol:base) of 1:1.5 and 1:1 for NaOH and DBU, respectively. The optical density of the samples was set to a value of 3 at λ_{pump} which corresponds to concentrations in the range of $(0.2-2.2) \cdot 10^{-2}$ mol/L. Since sufficient UV pulse energy at 418 nm was only yielded by SFM (sum of 775 and 908 nm, see section 3.2.1), probing with the sapphire WL was not possible.

The experimental time resolution of the setup using selected probe wavelengths was measured by cross correlations of the dyes BiBuQ (4,4'''-Bis-(2-butyloctyloxy)-p-quaterphenyl) in 1,4-dioxane[88] (for all nitrophenols) and pyridine 1 in ethanol[88] (for all phenolates) and found to be between 100 and 150 fs. Moreover, experimental time zero was independently determined by these cross correlation experiments at each probe wavelength. The time resolution of the transient setup was measured by cross correlations of the dye 7-diethylamino-3-thenoyl-coumarin (DETC) in ethanol[89, 90] to be always between 150 and 200 fs.

3.5.2 Photoinitiators

Benzoin (2-hydroxy-1,2-diphenylethanone, Bz, 98%) and 2-methyl-4'-(methylthio)-2-morpholinopropiophenone (MMMP, 98%) were purchased from Sigma Aldrich and used without further purification. 4-methyl-benzoin (4MB) was synthesized by the group of Prof. Dr. C. Barner-Kowollik, Institute for Chemical Technology and Polymer Chemistry, Karlsruhe Institute of Technology, according to reference 91. The samples were prepared as solutions in MIB. Time-resolved experiments were performed at 351 nm as well as 325 nm (λ_{pump}) using the transient spectrometer. Since all photoinitiators reveal (at least at 351 nm) very low extinction coefficients, an optical density of 3 would require high concentrations. In order to avoid effects like clustering or bimolecular quenching smaller OD values of 1.0 for MMMP and 0.6 for Bz and 4MB, respectively, were chosen. Due to low solubility of Bz and 4MB in MIB the same optical density as for MMMP could not be obtained. As mentioned in the previous section (3.5.1), the experimental time resolution was determined to be between 150 and 200 fs. Furthermore, WL generated data are only shown in the spectral range between 435 and 600 nm due to the aforementioned experimental artifact (section 3.5.1). This could recently be corrected by the implementation of an edge pass filter for 700 nm (Thorlabs) which drastically reduces the intensity of the dispersed WL around 700 nm, but was too late for measurements shown in this work. Adjustment of the continuously neutral density filter in front of the CCD camera now allows for detection of higher intensities in the spectral range between 600 and 650 nm.

3.6 Theoretical methods

DFT as well as TD-DFT calculations were performed using the TURBOMOLE V-6.3 program package.[92] Geometry optimizations in the lowest excited singlet and triplet states were computed with the BP86 functional and the def2-SV(P) basis set.[74, 75, 93-100] From the literature it is known that this combination leads to accurate minimum energies.[101] Excited state transitions[102-104] were calculated using the functional B3LYP[75, 76, 105] along with the basis set aug-cc-pVDZ[106] (for o-NP, Bz, 4MB and MMMP) and the basis set aug-cc-pVTZ[106] (for m- and p-NP and all nitrophenolates). Thereby, B3LYP is one of the most used functionals for computation of excitation energies.[70, 71]

For o-NP, minimum energy conical intersections (MECIs) between the first excited singlet and the ground state are calculated at the SA-2-CASSCF (state-averaged over two states) level of theory as implemented in the GAMESS-US (Version R1) employing a (7, 8)-CAS in combination with the 6-31G(d) basis set.[107-112] Since the first excited state is well characterized by a single electron LUMO–HOMO transition a smaller (4, 4)-CAS was employed for further PES scan. Thereby, a matrix of geometries was generated by varying the distance between the hydrogen atom of the hydroxyl group and the closest oxygen atom of the nitro group between 1.8 and 0.8 Å in steps of 0.2 Å. Furthermore, the dihedral angle between this oxygen atom and the phenyl plane was changed between 0° and 105° in steps of 15°. For each variation, the latter two internal degrees of freedom were fixed, whereas all others were allowed to relax. Finally, at the optimized geometries, energies were calculated with the MCQDPT (multi-configuration quasi-degenerate perturbation theory) method. [107, 113, 114]

For the $para$- and $meta$-nitrophenolates, calculations at the CASSCF level of theory were performed with the MOLCAS 8.0[115-117] and the GAUSSIAN 09[118] suites of programs in collaboration with C. García Iriepa (PhD student, Universidad de Alcalá, Spain) during a short stay at the Karlsruhe Institute of Technology. The minimum energy paths (MEPs) of the first excited singlet states were calculated using a (10, 8)-CAS in combination with the basis set 6-31G*.[109-112] As for o-NP, a matrix of geometries was created by varying the C–N stretching coordinate and the torsion angle around this bond.

4 Ultrafast dynamics of nitrophenols and corresponding nitrophenolates

4.1 Introduction

In this chapter, the dynamics of *ortho-*, *meta-* and *para*-nitrophenol as well as their corresponding phenolates are investigated by means of steady state and time-resolved TA spectroscopy combined with quantum chemical calculations. The results presented in this section belong to a manuscript to be submitted.[119] The nitrophenols (NPs) are characterized by comparable ICT transitions in the UV, but culminating in partly quite different dynamics. The molecular structures of the three constitutional isomers along with their pK_a values are shown in figure 4.1.

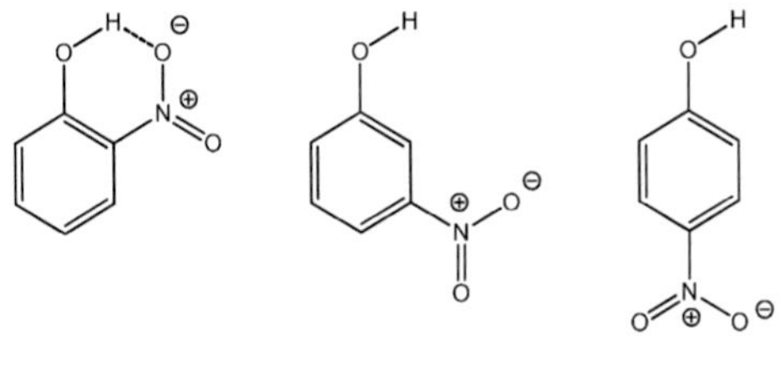

ortho (*o*-NP)	*meta* (*m*-NP)	*para* (*p*-NP)
pK_a = 7.23	pK_a = 8.36	pK_a = 7.15

Figure 4.1: Molecular structures and pK_a values[120] of *ortho-*, *meta-* and *para*-nitrophenol.

NPs are common environmental pollutants and belong to the group of volatile organic compounds (VOCs).[121] They are involved in chemical reactions in the atmosphere both in the gas and in the condensed phase (rainwater, fog, snow and cloud).[122] As identified by several experimental studies, formation of nitrated phenols is induced by photochemical reactions of precursor compounds, e.g. benzene, toluene, phenol and cresols.[123] In this regard, gas phase reaction between phenol, hydroxyl and nitrite radicals is supposed to yield p-NP as well as o-NP in the atmosphere during the night.[122] Furthermore, direct release of NPs is given by combustion of fossil fuels and by production of dye and explosive substances as well as pharmaceuticals.[122] Considering especially atmospheric reactions with hydroxyl radicals, o-NP is of particular interest.[124] Since hydroxyl radicals lead to oxidative decomposition of polluting trace substances (as e.g. VOCs) in the troposphere during the day, these radicals are commonly known as "detergents" of the atmosphere.[124] Further reaction products, however, are harmful photo-oxidants, e.g. ground-level ozone.[125, 126]

In this context, the photolysis of nitrous acid (HONO) in the UV (290-400 nm) was identified as the most important source of hydroxyl radicals.[127, 128] Evidence of HONO in the atmosphere was provided by Perner and Platt in 1979 for the first time.[129] In general, HONO can be released by direct emission[130] or heterogeneous reactions in the presence of nitrogen dioxide.[131] Due to further development of HONO detection methods unexpectedly high concentrations were detected indicating additional and previously unknown HONO sources.[132] To explain the observed high concentration values many experimental studies were performed which were able to identify three new possibilities of HONO formation.[133, 134] One of these is the gas phase photolysis of o-NP.[124, 135] After photoexcitation into the first excited singlet state the mechanism leading to release of HONO was proposed to proceed via an intramolecular proton transfer from the hydroxyl to the adjacent nitro group.[124] Theoretical studies suppose that this reaction occurs in the first excited triplet state.[136] A first indication of an excited state intramolecular proton transfer (ESIPT) reaction in o-NP was provided by detection of the tautomeric aci-nitro form by infrared spectroscopy in low-temperature argon matrices[137] and a resonance Raman investigation.[138] The aci-nitro form itself is supposed to be an unstable intermediate in the release of HONO.[124] Evidence by time-resolved experiments, however, has not been reported yet.

As discussed in previous studies[124, 125], NPs are small aromatic push-pull compounds which exhibit various ICT transitions upon excitation in the UV. Thus, these molecules serve as simple model compounds for investigation of ICT dynamics due to the absence of further functional groups except of the donor and acceptor moieties. Several photoproducts have been identified in aqueous solution, as e.g. catechols and resorcinols.[135, 139, 140] Thereby, high dependence on solvent environment, in particular pH value, and excitation wavelength has been found. Increased generation of photoproducts was observed after irradiation into electronic singlet states above S_1. Nevertheless, there are no detailed insights into the relaxation behavior of NPs yet. Triplet formation is suggested by ps time-resolved transient grating experiments conducted by Takezaki et al.[141] In summary, ISC is estimated to be ≤ 50 ps for all isomers, whereas the lifetime of the first excited triplet state T_1 is found to be 600 ps for m- and p-NP and 900 ps for o-NP. The longer triplet lifetime observed for the *ortho* isomer was attributed to its strong intramolecular hydrogen bond. Phosphorescence, however, was only detected at low temperatures (77 K) as well as in crystals for m- and p-NP.[142, 143]

In order to shed light on the dynamics of NPs, TA experiments in different solvents were performed, namely chloroform, 2-propanol and deionized water as well as n-hexane in the case of o-NP. The used solvents differ in their polarities and capabilities of hydrogen bonding (for dipole moments see table 4.1), and thus allow for investigation of solvent effects. Strong dependence of different assignment of the functional groups as well as influence of presence/absence of the intramolecular hydrogen bond was found resulting in common characteristic TA features for m- and p-NP. As a consequence, relaxation pathways of o-NP are discussed first (see section 4.2), whereas results of m- and p-NP are shown together thereafter (see section 4.3). Moreover, to study the influence of deprotonation on the excited state properties, further experiments were performed on the phenolate forms (see section 4.4). The deprotonated forms are stabilized by mesomeric effects which cause the weak acid character of NPs (for pK_a values[120] see figure 4.1). An overall summary of the relaxation pathways of protonated and deprotonated NPs is provided at the end of this chapter (see section 4.5).

Table 4.1: Dipole moments of employed solvents.[120]

solvent	Dipole moment /D
n-hexane	0
chloroform	1.01
2-propanol	1.66
water	1.85

4.2   Dynamics of *ortho*-nitrophenol (*o*-NP) after intramolecular charge transfer excitation

4.2.1   Steady state spectroscopy and calculated vertical transitions

Steady state spectra of *o*-NP are shown in figure 4.2 together with calculated oscillator strengths of vertical singlet-singlet transitions revealing A' symmetry. Since the oscillator strengths derived for singlet-singlet transitions with A" symmetry are smaller by several orders of magnitude, these excitations are attributable to transitions into dark singlet states. As a consequence, only the wavelengths of these transitions are marked in figure 4.2 together with the wavelengths of vertical singlet-triplet excitations. All calculations were performed at the TD-DFT level of theory as describe in section 3.6 and are related to the electronic ground state of *o*-NP, 1A' (S$_0$) (point group C$_S$, see section 4.2.3.1). The numerical values are listed in appendix B table B.7.

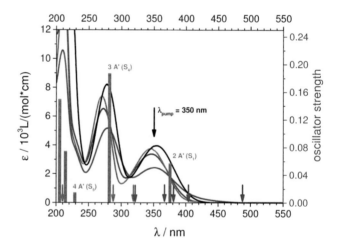

Figure 4.2: Measured absorption spectra of *o*-NP in *n*-hexane (grey), chloroform (black), 2-propanol (blue) and water (red). Calculated oscillator strengths of selected singlet-singlet transitions with A' symmetry are displayed as grey bars (right axis). Singlet-singlet transitions with A" symmetry are marked with grey and singlet-triplet transitions with blue arrows. The excitation wavelength of time-resolved experiments (λ_{pump}) was at 350 nm.

Considering figure 4.2, positions of the measured absorption bands agree well with the calculated vertical transitions into bright singlet states (A' symmetry) and exhibit only slight solvent dependence. The first absorption band is centered at 346 nm in *n*-hexane and 2-propanol, at 355 nm in chloroform as well as at 350 nm in water. These findings are in line with a previously recorded steady state spectrum in methanol, where the respective wavelength is found to be at 349 nm.[144] Since the overall spectral difference of the first maxima is approximately 9 nm, and thus small, the excitation wavelength (λ_{pump}) for all time-resolved experiments was chosen to be at 350 nm. The observed slight red-shift in polar solvents is most likely due to different extents of ground and excited state stabilization.[145] Similar extinction coefficients are found in *n*-hexane and chloroform, whereas smaller values are determined in 2-propanol and water. Table 4.2 contains the extinction coefficients obtained at the maximum wavelength (λ_{max}) of the first absorption band.

Table 4.2: Extinction coefficients obtained at maximum wavelength (λ_{max}) of the first absorption bands for *o*-**NP** in different solvents.

o-**NP**	*n*-hexane	chloroform	2-propanol	water
λ_{max}/nm	346	355	346	350
$\varepsilon/10^3$L/(mol·cm)	3.76	3.96	3.38	2.44

According to the TD-DFT calculations, excitation at 350 nm corresponds to transition into the first excited singlet state (2A' (S_1)). As inspection of molecular orbitals shows, excitation into this state is associated with a transition from the highest occupied (HOMO) into the lowest unoccupied molecular orbital (LUMO) and characterized by an intramolecular charge transfer from the hydroxyl to the adjacent nitro group. The involved orbitals are displayed in figure 4.3. The HOMO is featured by a π-MO revealing electron distribution over the benzene ring and the lone pair of the hydroxyl group. In contrast, the main contribution of the LUMO is given by π^*-MO implying the aromatic ring as well as the lone pairs of the nitro group. In summary, the first excited singlet state is described by $^1(\pi\pi)^*$ character. As can be seen from figure 4.2, several, nearly isoenergetic triplet states exist below S_1. For the sake of clarity, symmetry notation of excited states is neglected in the following.

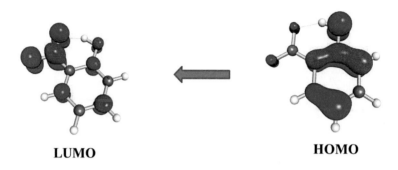

LUMO   **HOMO**

Figure 4.3: HOMO and LUMO orbitals for *o*-NP involved in the 2A' (S_1) ← 1A' (S_0) excitation.

4.2.2   Transient absorption spectroscopy

After ICT excitation into S_1, the excited state evolution of *o*-NP was monitored by selected time-delayed fs probe pulses in a wavelength range between 480 and 1100 nm. Since the employed probe wavelengths are outside the ground state absorption of *o*-NP, ground state-bleaching can be excluded on TA profiles. TA traces of *o*-NP are displayed in figure 4.4 and figure 4.5.

Since only a slight dependence on solvent is observed, all TA traces are discussed together. As one can see from figure 4.4 and 4.5, three spectral regions related to different probe wavelengths can be distinguished. The first region around 500 nm (subpanel a) is characterized by ESA, and thus positive ΔOD values, decaying on a sub-ps time scale. Weak residual absorption was observable in *n*-hexane (480 nm) and 2-propanol (480 and 500 nm) at long delay times within the investigated time window of 500 ps. Figure 4.6 provides a closer look at this feature in *n*-hexane; corresponding TA traces in 2-propanol are shown in appendix B, figure B.1. The observed remaining absorption highly depends on solvent and probe wavelengths and is probably too small to be detected in chloroform and water with the prevailing signal-to-noise ratio. The second subpanels (b), between 540 and 700 nm, are characterized by a superposition of ESA and SE within 0.5-1 ps. Thereby, changes in ΔOD occur on a wide range of probe wavelengths in the visible, but almost independent of solvent. In the third subpanels (c), only ultrafast SE was observed from 900 to 1100 nm in all solvents.

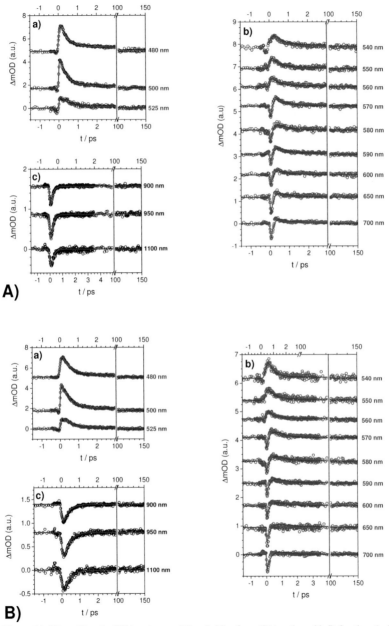

Figure 4.4: TA profiles of *o*-**NP** in *n*-**hexane (A)** and **chloroform (B)** together with fit functions (red lines) recorded at selected probe wavelengths as labeled. For explanation of the subpanels a)–c), see text. The pump wavelength was at 350 nm. Note that the plots are shifted along the ordinate for better clarity.

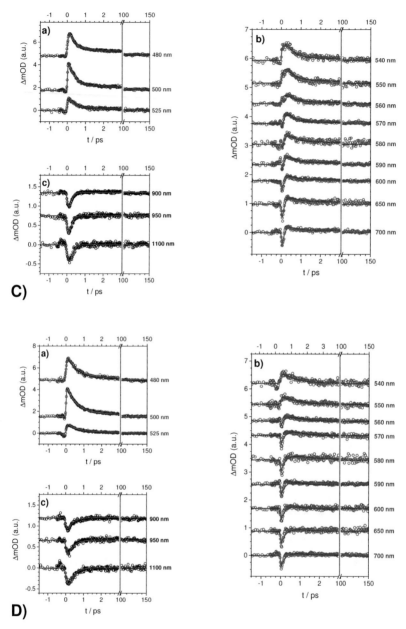

Figure 4.5: TA profiles of *o*-NP in **2-propanol** (**C**) and **water** (**D**) together with fit functions (red lines) recorded at selected probe wavelengths as labeled. For explanation of the subpanels a)–c), see text. The pump wavelength was at 350 nm. Note that the plots are shifted along the ordinate for better clarity.

In order to obtain a quantitative picture of the observed features, TA profiles of each subpanel are analyzed by fit functions as described in section 3.4 using a global fit routine. In the first subpanels a), functions best fit ESA around 500 nm with a sum of two exponentials (τ_1 and τ_2). Additionally, a third time constant τ_3 was required to model the remaining absorption for long delay times (only detectable in n-hexane and 2-propanol) associated with a long-lived excited species. Furthermore, monoexponential decay (τ_{ind}) was found for SE in the NIR (subpanels c)). All obtained time constants are listed in table 4.3. The errors are standard deviations from the global fitting procedure. Since τ_3 is beyond the experimental time window of 500 ps, this value is given as a lower boundary. For the intermediate probe wavelengths (subpanels b)), TA traces can be reckoned as superposition of ESA and SE. Hence, satisfying results of global fitting were achieved using a triexponential decay function which contains the previously received time constants τ_1, τ_2 and τ_{ind}. In this range of probe wavelengths, no residual absorption was detected for long delay times, and thus τ_3 is not significant. All fit functions are displayed as red lines along with the measured TA profiles in figure 4.4 and figure 4.5. Comparing different solvents, the values obtained for τ_{ind}, τ_1 and τ_2 are similar to each other (see table 4.3) except of τ_2 in water which is smaller by a factor of ~2. In addition, faster values by a factor of ~2 were found for τ_{ind} compared to τ_1. This behavior indicates two different effects: one causing SE in the visible up to NIR and the other leading to the dominant ESA around 500 nm.

Table 4.3: Time constants of o-NP obtained by global fit routine in four different solvents. Monoexponential decay (τ_{ind}) was found for SE between 900-1000 nm. Bi- and triexponential decays, respectively, were found for ESA between 480-500 nm. It is noted that τ_3 lies outside of the experimental time window of 500 ps.

o-NP		n-hexane	chloroform	2-propnaol	water
SE	τ_{ind}/ps	0.2 ± 0.1	0.3 ± 0.1	0.2 ± 0.1	0.3 ± 0.1
	τ_1/ps	0.4 ± 0.1	0.5 ± 0.1	0.4 ± 0.1	0.5 ± 0.1
ESA	τ_2/ps	9.2 ± 1.2	8.0 ± 0.9	9.8 ± 1.3	3.7 ± 0.3
	τ_3/ps	>500	---	>100	---

At this point it should be noted that the temporal as well as spectral range over which the negative TA values emerge is extraordinary for *o*-NP and raises a lot of questions. The promptness of the appearance of the SE likely points at a transition from the first excited to the ground state. However, SE in the NIR would imply an unusually high Stokes-shift compared to the excitation at 350 nm. One conceivable explanation for such a phenomenon could be that significant changes in the (electronic) molecular structure strongly influence the development of the PESs of the involved states. Therefore, it is indispensable to combine the results obtained from the TA experiments with quantum chemical calculations in order to get more insights into the excited states properties of *o*-NP.

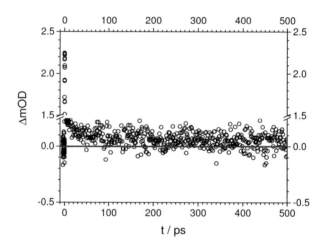

Figure 4.6: Weak residual absorption is observable for *o*-NP in *n*-hexane at an excitation wavelength of 350 nm and a probe wavelength of 480 nm on a hundreds of ps time scale.

4.2.3 Quantum chemical calculations

4.2.3.1 Density functional theory

Geometry optimizations of the ground state (S_0) as well as the first excited singlet (S_1) and triplet (T_1) state are performed at the (TD)-DFT level of theory (see section 3.6). All calculated Cartesian coordinates are listed in appendix B, table B.1, B.7 and B.13. The optimized structures are displayed in figure 4.7. The calculations indicate a planar structure of o-NP in S_0 and T_1 corresponding to the point group C_S. The values for OH···ONO and O–HONO distances were computed to be 1.609 and 1.021 Å in S_0 pointing to a strong intramolecular hydrogen bond in line with previous studies.[136, 146, 147] Contrary bond lengths are found for T_1 with 1.047 Å (OH–ONO) and 1.537 Å (O···HONO). Moreover, relative energies (E_{rel}) of the first five triplet excitations are calculated related to T_1. The values are summarized in table 4.4 together with their corresponding oscillator strengths.

Table 4.4: Calculated relative energies (E_{rel}) in eV and corresponding oscillator strengths of selected triplet transitions of o-**NP** related to T_1.

triplet states	E_{rel}/eV	oscillator strength
T_2	0.79	0.392
T_3	2.57	0.172
T_4	2.68	$1.53 \cdot 10^{-4}$
T_5	3.40	$1.02 \cdot 10^{-5}$
T_6	3.63	0.825

In view of the observed ultrafast SE in the TA experiments, excited state properties of the first excited singlet state (S_1) are also of particular interest. An analog attempt to optimize the geometry of this state, however, failed and no stable minimum structure could be found. The calculation indicates high structural changes in S_1 due to proton transfer from the hydroxyl to the nitro group accompanied by out-of-plane torsion of the newly formed HONO group. Thereby, close approach of the PES of S_1 and S_0 is predicted, resulting in a very small energy gap between both states. Figure 4.7 c) shows the last calculation step of the S_1 optimization which is characterized by a dihedral angle (ONCC, labeled by * in the structure) of ~54°. Since the failure of TD-DFT calculations in describing excited states which approach the ground state, and in general in describing conical intersections, is commonly known[80], multireference methods were employed to get further insights.

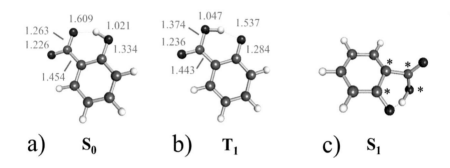

Figure 4.7: Optimized planar structures of *o*-**NP** for S_0 (**a**) and T_1 (**b**) together with selected distances in Å. A dihedral angle ONCC (marked by *) of ~54° is found for S_1 (**c**). For more details, see text above.

4.2.3.2 Multireference methods

In collaboration with T. J. A. Wolf and A. E. Boguslavskiy, the minimum energy conical intersection (MECI) between S_1 and S_0 was calculated at the CASSCF level of theory as described in section 3.6. In figure 4.8, the relative energy (E_{rel}, black dots) is plotted as a function of the HONO torsion angle and the H–O(NO) distance. The surfaces themselves were obtained by two-dimensional spline interpolations. The HONO torsion angle is defined as the difference between 180° minus the dihedral angle (H)CCNO(H) marked by * in the inserted molecular structure. Considering figure 4.8, only slight decrease in energy, beginning at the Franck-Condon point in S_1, is found by hydrogen transfer from the hydroxyl to the adjacent nitro group. The term "quasi-Franck-Condon point" is introduced because not all degrees of freedom were allowed to relax during energy calculations. The HONO torsion angle as well as the H–O(NO) distance were fixed at each point. In contrast, high changes of S_1 energy evolve by out-of-plane rotation of the HONO group and finally lead to the MECI with the ground state. The corresponding HONO torsion angle at the optimized MECI structure is calculated to be ~73° (Cartesian coordinates are listed in appendix B, table B.24). As displayed in figure 4.8, CI with the ground state can be reached by hydrogen shift to the nitro group combined with out-of-plane rotation of the HONO group via the two saddle points SP1 and SP2. Conclusions concerning the sequence of intramolecular nuclear motions (hydrogen transfer vs. torsion motion), however, have to be drawn with caution. Regarding different reduced masses of the involved atoms, hydrogen transfer presumably precedes rotation, but a more detailed answer can only be provided by detailed trajectory or wave packet calculations.

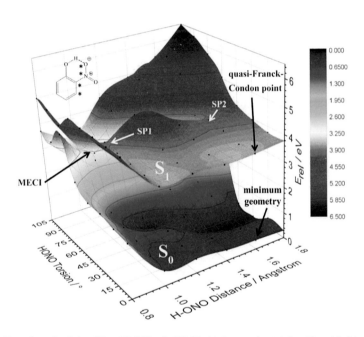

Figure 4.8: Three-dimensional plot of S_1 and S_0 PESs of *o*-**NP**. CI with the ground state is found by minimizing the H–O(NO) distance and increasing the HONO torsion angle. SP1 and SP2 are two saddle points towards the minimum energy conical intersection (MECI) with S_0. For more details, see text above.

4.2.4 Relaxation pathways of *ortho*-nitrophenol (*o*-NP)

4.2.4.1 Transient absorption profiles in the near infrared

After ICT excitation into S_1, SE occurs within 0.2-0.3 ps (nearly independent of solvent) extending over a wide range of probe wavelengths from the visible to the NIR. As already mentioned before, SE in the NIR would imply an unusually high Stokes-shift at an excitation wavelength of 350 nm, and thus must have another origin. Following these considerations the rapid appearance of SE points to an expeditious drop of the energy gap between the first excited singlet and the ground state. Considering section 4.2.3.1, such a behavior is already indicated by the performed geometry optimization of S_1 at the TD-DFT level of theory and further confirmed by the CASSCF calculations (see section 4.2.3.2). The latter reveal strong changes in the molecular structure in S_1 and a close approach of S_1 and S_0 while minimizing the (O)H–O(NO) distance and enlarging the HONO torsion angle. Hence, the entire relaxation process can be regarded as a combination of intramolecular excited state proton transfer coupled to the HONO rotation coordinate, and thus as a twisted ICT process finally leading to CI with S_0. Thereby, the ESIPT is most likely triggered by the charge transfer character of S_1. From figure 4.8, it can be concluded that reduction of the energy gap between S_1 and S_0 is more sensitive to the torsion mode than the hydrogen shift itself. Taking into account figure 4.3, high stabilization of the LUMO is suggested due to decoupling of donor and acceptor moieties along the HONO torsional coordinate.[35, 41]

Since the observed decay of SE is monoexponential, the associated time constant τ_{ind} is attributable to depopulation of S_1 induced by the probe pulse. In other words, the formation of the much discussed highly unstable intermediate, the *aci*-nitro isomer[124], is directly monitored by the time-resolved experiments. However, as τ_{ind} is very close to the experimental time resolution (~0.1 ps), the corresponding decay could be faster than this time constant may suggest. The proposed relaxation channel leading to SE is sketched in figure 4.9. Regarding the energetic changes between S_1 and S_0 along the suggested relaxation pathway, one might expect a red-shift of SE caused by wave packet motion from the initially excited Franck-Condon region toward the CI. For example, direct experimental evidence for such a shift is provided by sub-20-fs resolved TA experiments of the photoisomerization of retinal.[17] Even though the experimental time zero was independently determined by cross correlation experiments at each probe wavelength as described in section 3.5.1 (see appendix B, figure B.2 for reference experiments), evaluation of the TA data exhibits just an indication for a spectral shift. As demonstrated by the example of *n*-hexane in figure 4.10, the data do not provide an unambiguous dependence of the SE on the employed probe wavelengths. The ambiguity of the spectral shift is at least partly correlated to the intrinsic superimposition of SE and ESA extended over a wide spectral area. As a result, an alleged earlier decay of SE is observed, as e.g. seen at 600 and 700 nm in figure 4.10. Moreover, another reason for the unclear shift may be that the wave packet is broadening along the relaxation coordinate (ESIPT combined with out-of-plane rotation of the HONO group). For a closer look to selected TA traces in the other solvents the reader is referred to Appendix B, figure B.3. Furthermore, it has to be noted that the obtained time constant τ_{ind} is not necessarily identical to a time constant associated with depopulation of S_1 via conical intersection with the ground state, but might be smaller. Thus, τ_{ind} has to be reckoned as a lower boundary for the IC process.

Comparing to the literature, formation of an intermediate species triggered by ESIPT has also been detected for *ortho*-nitrobenzaldehyde[148] characterized by ultrafast fluorescence dynamic <100 fs. Moreover, ultrafast SE was observed for *para*-nitroaniline by Ernsting and coworkers using the pump-supercontinuum probe technique combined with semi-empirical quantum chemical calculations.[149, 150]

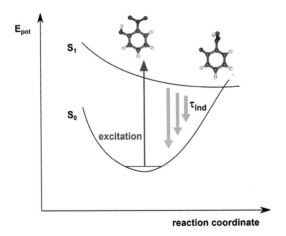

Figure 4.9: Part of the proposed relaxation pathway of *o*-**NP**: Excitation at 350 nm corresponds to ICT transition into S_1. Relaxation along the reaction coordinate (ESIPT coupled with out-of-plane rotation of the newly formed HONO group) leads to close approach of S_1 and S_0 PESs which is monitored by SE (green arrows, τ_{ind}). Thus, formation of the highly unstable *aci*-nitro form is detected.

In this regard, after excitation at 400 nm, SE occurs in a wavelength range between 430 and 700 nm within the first few hundreds of fs. This feature was assigned to IC into the ground state along a twisting coordinate of the nitro group. In contrast to other aromatic compounds exhibiting twisted ICT and partly also dual fluorescence,[33, 35, 42] as e.g. 4-(N,N-dimethylamino)-benzonitrile, o-NP obviously have no *stable* twisted form in its ICT state. Considering especially ESIPT reactions, typical time constants are on a 10 fs time scale, as e.g. found for 2-(2'-hydroxyphenyl)benzothiazole,[151, 152] being well in line with the findings for o-NP. Nonetheless, one would expect an influence on related time constants due to different polarities or capabilities for hydrogen bonding of the solvent, e.g. observed for 2-(2'-hydroxyphenyl)-benzothiazole[153] and 2-(2'-hydroxy-5'-methylphenyl)-benzotriazole.[154] In these cases, slower time constants or even lack of ESIPT reactions in hydrogen bonding solvents indicate that interaction with solvent molecules highly affect the intramolecular proton shift. For o-NP, however, no significant solvent effect on the obtained time constant τ_{ind} (ESIPT coupled with out-of-plane rotation of the HONO group) is observed. This result points to a barrierless reaction coordinate which is supported by the performed CASSCF calculations. According to literature, a barrierless ESIPT reaction is e.g. found for paeonol characterized by its independence on solvent environment.[155]

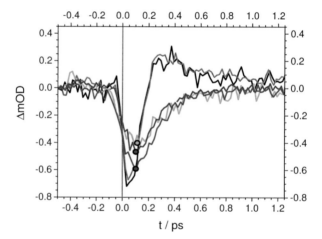

Figure 4.10: Selected TA profiles of *o*-NP in *n*-hexane at 650 nm (black), 700 nm (grey), 900 nm (blue), 950 nm (red) and 1100 nm (green). As indicated by the filled circles, only an indication for a red-shift of the SE is given. For more details, see text.

4.2.4.2 Transient absorption profiles in the visible region

Considering figure 4.6 and figure B.1 in appendix B, residual TA for long delay times was recorded in *n*-hexane at 480 nm, and in 2-propanol at 480 nm and 500 nm, respectively, indicating a long-lived excited species. This observation agrees well with the calculated triplet excitation energy at about 2.57 eV related to T_1 (table 4.4) which corresponds to a probe wavelength of 482.4 nm. Further triplet transitions (at higher or lower excitation energies) with non-negligible oscillator strengths very likely have no effect on TA experiments because they lie outside the range of used probe wavelengths. Consequently, the time constant τ_3 assigned to the remaining TA can be associated with absorption of the first excited triplet state T_1, and thus with the lifetime of this state. Comparing the obtained time constants with a previous investigation of *o*-NP in benzene by Takezaki *et al.* (900 ps), faster values for the lifetime of T_1 were found in 2-propanol (>100 ps) and *n*-hexane (>500 ps). Differences between these values can be realized by consideration of the strong solvent dependence of τ_3. Since the interaction between excited *o*-NP and the solvent environment is much stronger in polar solvents, it is reasonable that the energy transfer to polar solvent molecules is also much faster and that the shortest time constant was found in 2-propanol.

As calculated at the TD-DFT level of theory, two very close lying triplet states, namely T_3 and T_2, are below S_1 (see figure 4.2). Taking into account the $^1(\pi\pi)^*$ character of the first excited singlet state of *o*-NP, such states can easily undergo ISC processes into triplet states with $^3(n\pi)^*$ character as predicted by the El-Sayed selection rules.[156, 157] Inspection of molecular orbitals, however, indicates $^3(\pi\pi)^*$ character for T_2 and T_3. Nevertheless, $^3(n\pi)^*$ character is found for T_4 whose energy lies slightly above the calculated S_1 transition. Following these considerations, the time constant τ_1 is attributable to depopulation of S_1 via both ISC and IC and has to be expressed as $\tau_1 = (\tau_{IC} \cdot \tau_{ISC})/(\tau_{IC} + \tau_{ISC})$. As a result, τ_1 can be only reckoned as a lower limit to the underling ISC process. Moreover, τ_2 is then related to vibrational relaxation in the upper triplet manifold as well as IC into the lowest lying triplet state T_1 and vibrational relaxation therein. Analogously to τ_1, τ_2 is nearly independent of solvent (table 4.3). One exception, however, is the value obtained in water which is by a factor of ~2 lower compared to the other solvents. In this case, the energy transfer is likely facilitated by the higher polarity and proticity of water in comparison to the other solvents (for dipole moments see table 4.1). The supposed relaxation pathway via ISC is sketched in figure 4.11. Further contributions on TA traces due to HGSA cannot fully be excluded as a result of competitive relaxation processes. Taking into account the performed geometry optimization of T_1 at the TD-DFT level of theory (section 4.3.2.1), relaxation into T_1 is linked to a proton transfer, and thus formation of the *aci*-nitro form.

In comparison to literature[141], the obtained values of 0.4-0.5 ps for τ_1 are far below the value estimated by Takezaki *et al.* for *o*-NP in benzene ($\leq$50 ps). Time constants on a sub-ps time scale, however, have been found for several nitrated polycyclic aromatic compounds, e.g. nitronaphthalene[158], as well as for benzene.[159] In analogy to the findings for *o*-NP, nearly isoenergetic triplet states below S_1, and thus very low energy gaps between S_1 and upper triplet states, are a typical feature of such molecules. Comparing the findings for *o*-NP with those for *para*-nitroaniline[34], a rapid and efficient ISC into T_1 was found in a range of $\leq$0.8-10 ps dependent on solvent.

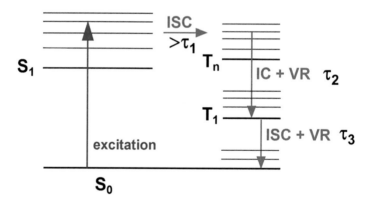

Figure 4.11: Part of the proposed relaxation pathways of *o*-**NP**: Excitation at 350 nm corresponds to ICT transition into S_1. Depopulation of S_1 via ultrafast ISC (> τ_1), IC accompanied by vibrational relaxation (VR) in the triplet manifold (τ_2) and depopulation of T_1 (τ_3) is observed.

4.2.5 Conclusion

In summary, it can be stated that the relaxation dynamics of o-NP in solution are mainly characterized by two aspects after ICT excitation into S_1:

1) As revealed by CASSCF calculations, a close approach of S_1 and S_0 PESs due to ESIPT combined with out-of-plane rotation of the newly formed HONO group is found toward CI with the ground state. The ESIPT reaction itself is most likely triggered by the ICT character of S_1. Depopulation of S_1 via this relaxation pathway is observed by SE in the visible region up to the NIR within 0.2-0.3 ps dependent on solvent.

2) Ultrafast ISC facilitated by nearly isoenergetic character of triplet states below S_1, as well as by the El-Sayed selection rules[156, 157], is identified as a competitive relaxation channel regarding the depopulation of S_1. A lower limit for ISC is found to be at 0.4-0.5 ps dependent on solvent.

The present work provides a time-resolved study of the structural changes in o-NP after ICT in the UV in solution for the first time. Furthermore, additionally performed time-resolved experiments in the gas phase by means of photoelectron spectroscopy indicate an analog relaxation behavior.[119] Evidence of the unstable *aci*-nitro isomer, which forms a cornerstone for HONO generation in the atmospheric o-NP photolysis, is given in both the liquid and gas phase. Due to the observed ultrafast singlet dynamics the much discussed HONO release is supposed to take place on the triplet manifold.[124, 136]

In order to answer the question how different assignments of the two functional groups in nitrophenol and the absence of intramolecular hydrogen bond influence the excited state properties, TA experiments of the isomers m- and p- NP were performed and are discussed in the following part.

4.3 Dynamics of *meta*- and *para*-nitrophenol (*m*- and *p*-NP) after intramolecular charge transfer excitation

4.3.1 Steady state spectroscopy and calculated vertical transitions

Steady state spectra of *m*- and *p*-NP are shown in figure 4.12 a) and b), respectively, along with calculated vertical single-singlet and singlet-triplet transitions. As mentioned in section 3.5.1, *n*-hexane could not be employed due to too low solubility of *m*- and *p*-NP. In analogy to *o*-NP, both isomers exhibit a planar minimum geometry in their electronic ground state corresponding to C_S symmetry. Likewise, singlet-singlet excitations with A' symmetry can be assigned to transitions into bright single states, whereas singlet-singlet excitations with A" symmetry corresponds to transitions into dark singlet states. For the sake of clarity, only the lowest three singlet states with A' symmetry are explicitly labeled in figure 4.12. All calculations are related to the ground state 1A' (S_0). For a summary of the numerical values and the computed coordinates of the optimized ground state structures, the reader is referred to appendix B table B.17, B.18, B.2 and B.3.

From figure 4.12 a) it can be realized that the first absorption band of *m*-NP is centered at 329 nm (chloroform) and 331 nm (2-propanol and water) revealing only slight solvent dependence. Considering the calculations, excitation at 330 nm corresponds to a transition into the first excited bright singlet state, 2A' (S_1). In contrast, the first absorption band of *p*-NP is centered at 311 nm (chloroform), 314 nm (2-propanol) and 317 nm (water) (see figure 4.12 b)). It is obvious that the first absorption band is in reasonable agreement with the calculated position of the vertical transitions into the first excited bright singlet state, 2A' (S_2). As inspection of the molecular orbitals indicates, excitation into these states is attributable to transition from the HOMO into the LUMO. The involved orbitals are displayed in figure 4.13 a) and b) for *m*- and *p*-NP, respectively. Analogously to *o*-NP, the HOMO of both isomers is characterized by a π-MO revealing electron distribution over the benzene ring and the lone pair of the hydroxyl group. In contrast the LUMO is described by a π*-MO implying the benzene ring as well as the lone pairs of the nitro group. Consequently, it can be stated that the LUMO–HOMO transition is well defined by an intramolecular charge transfer from the hydroxyl to the nitro group into a $^1(\pi\pi^*)$ state. Compared to *o*- and *m*-NP, TD-DFT calculations reveal a dark singlet state (1A" (S_1)) below the first bright one (2A' (S_2)) for *p*-NP which can be described as a LUMO–HOMO-6 transition.

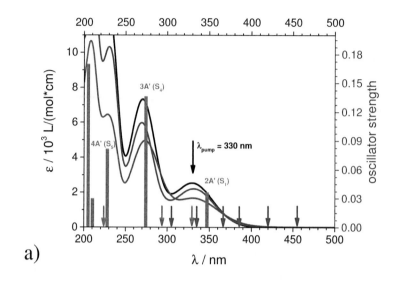

a)

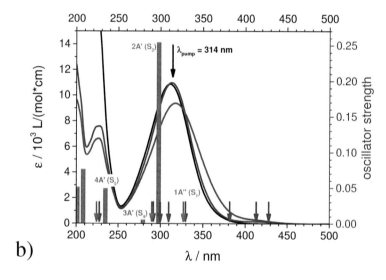

b)

Figure 4.12: Measured absorption spectra of *m*-NP **(a)** and *p*-NP **(b)** in chloroform (black), 2-propanol (blue) and water (red). Calculated oscillator strengths of selected singlet-singlet transitions with A' symmetry are displayed as grey bars (right axis). Singlet-singlet transitions with A" symmetry are marked with grey and singlet-triplet transitions with blue arrows. The excitation wavelengths (λ_{pump}) of time-resolved experiments were at 330 nm (*m*-NP) and 314 nm (*p*-NP).

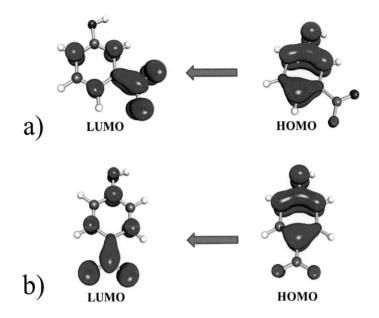

Figure 4.13: LUMO–HOMO orbitals of *m*-NP **(a)** and *p*-NP **(b)** involved in the 2A' (S$_1$) ← 1A' (S$_0$) excitation (*m*-NP) and 2A' (S$_2$) ← 1A' (S$_0$) excitation (*p*-NP).

Comparing the spectra of *m*- and *p*-NP, a shift to lower wavelengths is observable for the *para* isomer. This indicates higher coupling between donor and acceptor moieties caused by different topology and emerging resonance structures which is known from several studies of donor-acceptor anions containing nitro groups.[145] Furthermore, a slight red-shift is found for solvents with higher electric dipole moments and increasing capability of hydrogen bonding (water > 2-propanol > chloroform, see table 4.1) which seems to be more significant for *p*-NP. The solvatochromic shift is likely caused by different extents of ground and excited states stabilization.[145] In addition, oscillator strengths are directly linked to the coupling between donor and acceptor groups revealing higher values for stronger donor-acceptor interaction.[145, 160] Considering the extinction coefficients of the first absorption bands, larger values (by a factor of 4.3 to 5.6 depending on solvent) are found for *p*-NP in line with the presumed higher donor-acceptor coupling in comparison to *m*-NP. Values obtained at maximum wavelength (λ_{max}) of the first absorption band are listed in table 4.5. As mentioned above, the employed excitation wavelengths correspond to a transition into the first bright singlet state (2A') for both isomers. Moreover, TD-DFT calculations show several (partly nearly isoenergetic) triplet states below these excited singlet states. In the interest of simplification, symmetry notation of the excited states is neglected in the following sections.

Table 4.5: Extinction coefficients obtained at maximum wavelength (λ_{max}) of the first absorption bands for *m*- and *p*-**NP** in different solvents.

nitrophenol		chloroform	2-propanol	water
m-**NP**	λ_{max}/nm	329	331	331
	$\varepsilon/10^3$L/(mol·cm)	2.54	2.21	1.67
p-**NP**	λ_{max}/nm	311	314	317
	$\varepsilon/10^3$L/(mol·cm)	10.88	10.98	9.38

4.3.2 Transient absorption spectroscopy in solution

After ICT excitation into S_1 (*m*-NP) and S_2 (*p*-NP) the temporal evolution of the excited state depopulation was monitored by selected probe wavelengths (480-950 nm) as well as a sapphire WL continuum (see section 3.5.1). The two notions TA (sw/single wavelength) and TA (wl/white light) are introduced in order to distinguish between the different employed experimental methods, whenever it is necessary.

4.3.2.1 *para*-nitrophenol (*p*-NP)

For direct comparison, the performed TA (sw) and TA (wl) profiles are shown in figure 4.14 on the left and right hand side, respectively (a) chloroform, b) 2-propanol, c) water). Since the observed features show a very strong solvent dependence, TA traces in chloroform and 2-propanol are discussed together, whereas profiles recorded in water are examined separately. As can be seen from figure 4.14 a) and b), probing yields positive ΔOD values due to ESA.

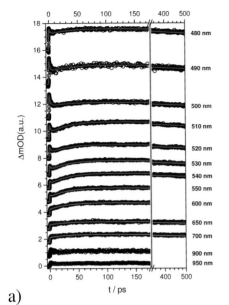

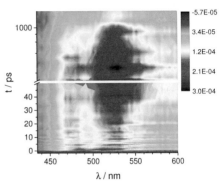

a)

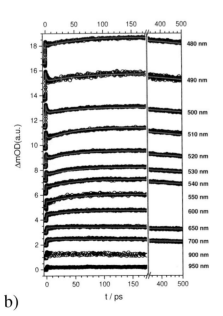

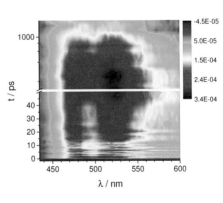

b)

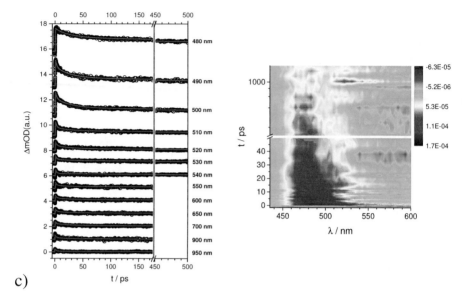

Figure 4.14: Comparison of TA (sw) and TA (wl) profiles of *p*-NP in **chloroform (a)**, **2-propanol (b)** and **water (c)** at a pump wavelength of 314 nm. Right hand side: TA (sw) profiles together with fit functions (red lines) recorded at selected probe wavelengths as labeled. Note that the plots are shifted along the ordinate for better clarity. Left hand side: TA (wl) profiles recorded using a white light continuum between 435 and 600 nm.

Thereafter, ΔOD is characterized by a fast decay within 5 ps after time zero followed by a rise within hundreds of ps, highly dependent on λ_{probe}. As inspection of TA (wl) directly shows, the observed dynamics (decay as well as rise) are related to two different absorption bands centered around 475 and 525 nm, respectively. Obviously, the band around 475 nm exhibits a higher intensity in 2-propanol compared to chloroform.

For a quantitative inspection of the observed features, the data obtained under both experimental conditions are analyzed by fit functions described in section 3.4. As a result of several tests, functions best fit TA with a sum of three exponentials. A Global fit routine over the entire probe wavelength range, however, does not achieve satisfying results. Consequently, the two observed absorption bands have to be analyzed separately by global fitting. For TA (wl) the spectral region between 459 and 498 nm as well as 501 and 549 nm are evaluated; for TA (sw) the traces 480-500 nm and 510-700 nm, respectively. Obtained fit functions are displayed as red lines together with the measured TA (sw) traces in the left part of figure 4.14. Since the intensity as well as the signal-to-noise ratio of the TA (sw) profiles at 900 and 950 nm is very low, these traces are not included in the global fit routine (for a closer look at both TA profiles see appendix B figure B.4). Corresponding decay associated difference spectra (DADS) of TA (wl) are shown and discussed later (see section 4.3.4.2). The time constants derived for both experiments are summarized in table 4.6. The given errors are standard deviations derived from the fitting procedure. If the time constants are beyond the experimental time windows of 500 ps (TA (sw)) and 1800 ps (TA (wl)), the time windows are given instead as a lower boundary. The first time constant τ_1 can be assigned to the observed decay of ΔOD within 5 ps after time zero. The second time constant τ_2 is attributable to the following signal rise resulting in a maximum ΔOD value around 110 ps in chloroform and 150 ps in 2-propanol. The decay of TA beyond the experimental time windows is related to τ_3.

Comparing the constants obtained for TA (sw) and TA (wl), the values generally agree very well. Differences, however, are found for τ_1 in chloroform ($\Delta\tau_1 = 2.1$ ps around 475 nm and $\Delta\tau_1 = 1.1$ ps around 525 nm) and for τ_2 in 2-propanol ($\Delta\tau_2 = 27.7$ ps around 475 nm and $\Delta\tau_2 = 22.3$ ps around 525 nm). As mentioned in section 3.4, experimental conditions of the employed setups differed in some aspects which can lead to the observed discrepancies. In this regard, higher intensity of ΔOD for TA (sw), and thus in many cases better signal-to-noise ratio, likely induces deviations in time constants associated with fast changes in ΔOD, as e.g. observed for the initial signal decay (τ_1) in chloroform. Hence, the obtained τ_1 value from TA (sw) experiments has a greater significance. Considering the experimental time windows of 500 ps and 1800 ps, respectively, determination of τ_2 is presumably influenced by changing τ_3 (as in the case of τ_2 in 2-propanol). In this case, determination of τ_2 is more precise for TA (wl) as for TA (sw). Furthermore, global analysis of different probe wavelength regions and numbers of TA traces may cause slight inconsistencies between time constants obtained under both experimental conditions. These aspects, however, are not included in the standard deviations derived from the fitting procedure. Based on the same experimental conditions, time constants determined around 475 and 525 nm exhibit similar values for τ_1, whereas τ_2 reveal higher values around 475 nm. In summary, it can be stated that an analog relaxation behavior is observable for *p*-NP in 2-propanol and chloroform after excitation into S_2 manifested in a fast decay and additional growth of ΔOD on the time scale of hundreds of ps.

After excitation of p-NP in water (figure 4.14 c)), probing yields positive ΔOD values due to ESA which decay on a ps time scale resulting in a constant long-time absorption. In contrast to chloroform and 2-propanol, ΔOD does not increase again after the initial decay around time zero in water. Appearance of two distinguishable absorption bands centered approximately at 475 nm and 507 nm is weakly indicated in TA (wl) (right part of figure 4.14 c)). Both bands are very close and seem to be superimposed. In addition, a slight shift to the blue is observable within ~30 ps. Since the band around 507 nm reveals low intensity, and thus rarely influences ΔOD, quantitative analysis of TA (wl) leads to three similar time constants for the observed absorption bands. As a result, the entire range between 465 and 525 nm of TA (wl), where the maximum intensity of ESA is observed, can be analyzed by a global fit function using three exponentials. Corresponding DADS of TA (wl) are shown and discussed later (see section 4.3.5). In a similar manner TA (sw) can be evaluated using a sum of three exponentials. The derived fit functions are depicted as red lines together with the TA (sw) profiles in figure 4.14 c) in the left part. Obtained time constants are listed in table 4.6. The third time constant τ_3 models the remaining absorption for long delay times, whereas τ_1 and τ_2 are associated to the initial decay after time zero and additional dynamics on the time scale of tens of ps. As can be seen from table 4.6, the values obtained under both experimental conditions are in very good agreement.

It can be noted that different relaxation pathways likely exist after excitation of p-NP leading to a decay of ΔOD in water and an additional increase within hundreds of ps in 2-propanol and chloroform. Thereby, a trend of ΔOD is found in the order of chloroform, 2-propanol and water or, in other words, in the order of increasing electric dipole moments of the employed solvents (see table 4.1): Intensity of the absorption band around 525 nm is decreasing, whereas intensity of the band around 475 nm is increasing.

Table 4.6: Time constants of *p*-**NP** obtained by global fit routine for TA (sw) and TA (wl) in chloroform, 2-propanol and water. The two distinguishable absorption bands are analyzed separately for chloroform and 2-propanol. The wavelength range is noted. The given errors are standard deviations derived from the fitting procedure.

solvent		τ_1/ps	τ_2/ps	τ_3/ps
chloroform	TA (sw)			
	480-500 nm	2.1 ± 0.1	46.2 ± 2.3	$\gg 500$
	510-650 nm	2.8 ± 0.1	35.0 ± 0.4	$\gg 500$
	TA (wl)			
	459-498 nm	4.2 ± 0.3	56.8 ± 5.2	> 1800
	501-549 nm	3.9 ± 0.3	41.9 ± 3.4	> 1800
2-propanol	TA (sw)			
	480-500 nm	1.8 ± 0.1	97.2 ± 3.7	$\gg 500$
	510-650 nm	1.9 ± 0.1	64.8 ± 0.9	$\gg 500$
	TA (wl)			
	459-498 nm	1.9 ± 0.1	69.5 ± 4.2	> 1800
	501-549 nm	2.2 ± 0.1	42.5 ± 1.8	> 1800
water	TA (sw)	2.5 ± 0.1	27.6 ± 0.6	$\gg 500$
	TA (wl)	2.3 ± 0.2	30.8 ± 2.2	> 1800

4.3.2.2 *meta*-nitrophenol (*m*-NP)

TA profiles probing the samples using selected probe wavelengths (between 480 and 950 nm) as well as a white light continuum (435-600 nm) are displayed in figure 4.15 on the left and right hand side, respectively (a) chloroform, b) 2-propanol, c) water). As observed for *p*-NP, ΔOD of *m*-NP is characterized by similar features in chloroform and 2-propanol, whereas traces in water exhibit a different behavior. In the following, we will first focus on TA in chloroform and 2-propanol; TA in water will be discussed afterwards.

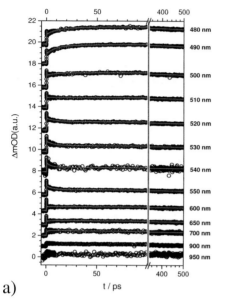

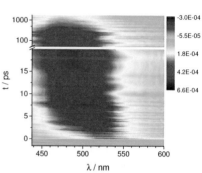

a)

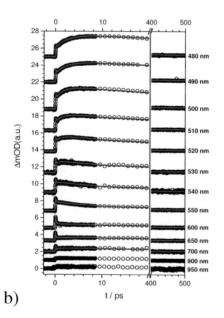

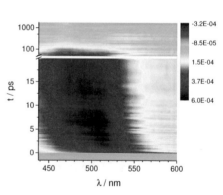

b)

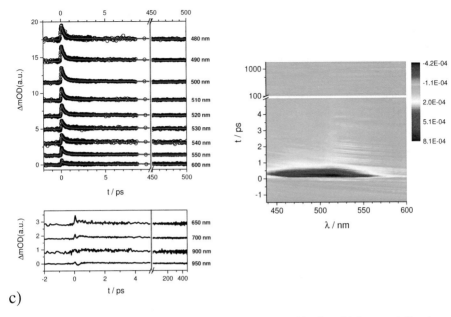

c)

Figure 4.15: Comparison of TA (sw) and TA (wl) profiles of *m*-**NP** in **chloroform (a)**, **2-propanol (b)** and **water (c)** at a pump wavelength of 330 nm. Right side: TA (sw) profiles together with fit functions (red lines) recorded at selected probe wavelengths as labeled. TA (sw) profils in water (c) between 650 and 950 nm could not be analyzed. Note that the plots are shifted along the ordinate for better clarity. Left side: TA (wl) profiles recorded using a white light continuum between 435 and 600 nm.

Figure 4.15 a) and b) shows that probing yields positive ΔOD values due to ESA after excitation into S_1. The TA profiles are defined by an ultrafast decay within the first ps after time zero and a subsequent signal growth on the ps time scale dependent on λ_{probe}. This increase appears in the blue probe wavelengths range and completely disappears for higher probe wavelengths up to the NIR (see TA (sw), left side of figure 4.15). As one can directly see from white light generated spectra, a spectral shift to the blue is observable within a pump probe delay of 5-10 ps in both solvents.

Quantitative analyses of the TA profiles in chloroform achieve best results with a sum of four exponentials. In the case of TA (wl) the entire range between 450 nm and 531 nm is evaluated. For TA (sw), traces recorded at 900 and 950 nm are excluded due to low intensity and signal-to-noise ratio (for a closer look at these traces see appendix B figure B.5). Since the decay of the intermediate signal growth is much faster in 2-propanol than in chloroform, a further time constant is required for characterizing the spectra recorded in 2-propanol for long delay times. For TA (wl), the spectral region between 450 and 549 nm is analyzed; as has already been mentioned before, TA (sw) traces at 900 and 950 nm are not included in the fit routine. Fit functions of TA (sw) are displayed as red lines together with the measured traces on the left side of figure 4.15; corresponding DADS spectra of TA (wl) are shown and discussed later (see section 4.3.4.2). All time constants are obtained using a global fitting routine except of τ_2 which is optimized separately for each λ_{probe}. The values are listed in table 4.7. If the time constants are beyond the values of experimental time resolution (~0.1 ps), this value is given instead as an upper boundary. The first time constant τ_1 is attributable to the ultrafast decay observed within the first ps after time zero. τ_2 and τ_3 are related to the following signal growth. Thereby, different values of τ_2 are required depending on λ_{probe} to model the observed spectral shift to the blue. Maximum ΔOD values are found to be at ~10 ps in 2-propanol and ~120 ps in chloroform. The signal decay for long delay times is related to τ_4 and τ_5 (only in 2-propanol), respectively.

As can be seen from table 4.7, time constants determined under both experimental conditions are in good accordance. One major exception, however, is τ_3 in chloroform ($\Delta\tau_3 = 114$ ps) which needs a closer examination. Figure 4.16 displays the direct comparison of TA (sw) and TA (wl) at 480 and 490 nm where the intermediate increase of ΔOD exhibits the highest intensity. The signal-to-noise ratio of both experiments is comparable, whereas ΔOD is approximately three times lower for TA (wl). Comparing the relative intensity of the two selected traces under the same experimental conditions, differences can occur due to various pump probe overlap by aligning the setup using selected probe wavelengths. After the initial decay TA (sw) rises within 100 ps and further decays within several hundreds of ps showing a flat descent of ΔOD. In contrast, intensity of TA (wl) slightly collapses between 30 and 200 ps which leads to an alleged faster value of τ_3. The spectra were recorded with different step size of data point collection and higher data acquisition was performed for pump probe experiments using selected probe wavelengths. Thus, TA (wl) shows a random error in τ_3 due to experimental conditions and it is obvious that the error deviation derived from fit routine is not realistic. Considering τ_3 obtained from TA (sw), a more adequate error bar is estimated to be ≥ 100 ps for τ_3 TA (wl). In summary, similar relaxation behavior is observed for m-NP in chloroform and 2-propanol which reveals faster time constants in 2-propanol.

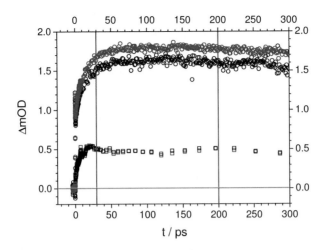

Figure 4.16: Comparison between **TA (sw, circles)** and **TA (wl, rectangles)** of *m*-**NP** at the two selected probe wavelengths 480 nm (black) and 490 nm (blue). The excitation wavelength was at 330 nm. Intensity of TA (wl) slightly collapses between 30 ps and 200 ps (labeled by red lines) leading to an alleged faster value of τ_3 derived from the white light generated spectra.

As displayed in figure 4.15 c), probing of *m*-NP in water yields positive ΔOD values due to ESA. The dynamics are mainly dominated by a fast decay within tens of ps. Moreover, no remaining absorption is observable for long delay times. Considering TA (sw) (figure 4.15 c) left side), ESA is superimposed by SE for probing beyond 600 nm. This interference leads to a nearly complete suppression of TA at 900 nm and negative ΔOD values at 950 nm. Since these features are outside of the accessible wavelength range of the generated sapphire white light, they cannot be detected for TA (wl).

Global analyses of TA (sw) below 650 nm, and of TA (wl) between 450 and 550 nm, results in two time constants which are listed in table 4.7. The obtained fit functions are shown as red lines together with TA (sw) profiles on the left side of figure 4.15; corresponding DADS are discussed later (see section 4.3.5). Both time constants, τ_1 and τ_2, are assigned to the fast decay within tens of ps and show very good agreement between the different employed experimental methods. Since the superimposed TA (sw) traces between 650 nm and 950 nm reveal very low intensities (nearly extinguished around 900 nm), quantitative evaluation is not possible. Comparing the obtained results for m- and p-NP, the dynamics of both isomers are quite complex, especially due to high solvent dependence. The following section aims at the recapitulation of the main aspects and includes further investigation of the appearance of SE in water.

Table 4.7: Time constants of m-**NP** obtained by global fit routine for TA (sw) and TA (wl) in chloroform, 2-propanol and water. τ_2 is optimized separately for each λ_{probe} in chloroform and 2-propanol. The given errors are standard deviations derived from the fitting procedure. For τ_3, the error is additionally estimated to be ≥ 100 ps. For more details, see text above.

solvent		τ_1/ps	τ_2/ps	τ_3/ps	τ_4/ps	τ_5/ps
chloroform	TA (sw)	≤ 0.1	(1.2-12.3) $\pm$ (0.3-0.5)	122.5 ± 10.0	$\gg 500$	---
	TA (wl)	≤ 0.1	(1.3-8.2) $\pm$ (0.3-1.0)	18.5 ± 1.0 (estimated error $\geq$ 100 ps)	> 1800	---
2-propanol	TA (sw)	≤ 0.1	(0.2-1.7) $\pm$ (0.1-0.5)	5.8 ± 0.2	59.2 ± 0.4	$\gg 500$
	TA (wl)	0.2 ± 0.1	(0.2-1.7) $\pm$ (0.1-0.7)	8.3 ± 0.5	64.6 ± 1.0	>1800
water	TA (sw)	2.5 ± 0.1	27.6 ± 0.6	---	---	---
	TA (wl)	2.3 ± 0.2	30.8 ± 2.2	---	---	---

4.3.3 Occurrence of stimulated emission in water

Reviewing the previous sections, TA of the isomers *m*- and *p*-NP reveal similar behavior in chloroform and 2-propanol mainly characterized by an increase of ΔOD between 10 ps and 150 ps dependent on solvent. In this regard, a clear distinguishable mark is the occurrence of two spectrally separable absorption bands for *p*-NP, whereas only one broad band is observable for *m*-NP. In contrast, diverse TA traces are found in water represented by decay of ΔOD. Hence, it is obvious that relaxation pathways in water differ from those in chloroform and 2-propanol. One remarkable point is the appearance of SE in the NIR observed for *m*-NP in water. By looking at this feature, high analogy to the dynamics of *o*-NP (see section 4.2) is evident. In order to get deeper insights into the dynamics appearing in water and to answer the question why SE is only observable in the case of *m*-NP, pump-probe experiments at a further excitation wavelength (323 nm) were performed (figure 4.17 a) *m*-NP, b) *p*-NP). In doing so, both isomers were probed at two selected wavelengths (600 nm (black) and 900 nm (blue)). As can be seen, SE is found for both molecules at different probe wavelengths. SE occurs at 900 nm for *m*-NP which is in good agreement with the previous results obtained for excitation at 330 nm, whereas SE is observable at 600 nm for *p*-NP, but not detectable upon excitation at 314 nm.

The following sections provide a detailed examination of all features and results described above. According to the observed solvent dependence the relaxation pathways of *m*- and *p*-NP are discussed together in i) chloroform and 2-propanol and ii) water.

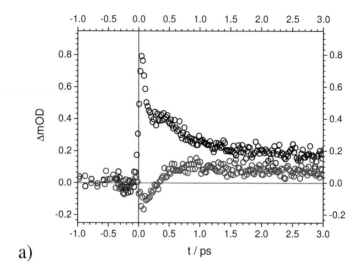

a)

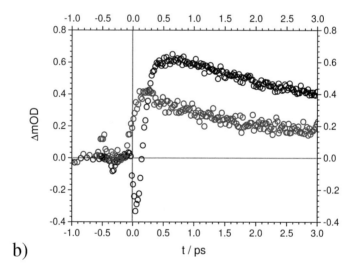

b)

Figure 4.17: TA of *m*-**NP (a)** and *p*-**NP (b)** in **water** at a pump wavelength of 323 nm and a probe wavelength of 600 nm (black) and 900 nm (blue), respectively. SE is observable for both isomers.

4.3.4 Dynamics in chloroform and 2-propanol

4.3.4.1 Photodegradation of nitrated polycyclic aromatic hydrocarbons

Taking into account TA traces of *p*- and *m*-NP in chloroform and 2-propanol (figure 4.14 and 4.15 a) and b)), remaining ESA on ps up to ns time scale points to formation of long-lived excited species, e.g. triplet states. Since the intermediate rise and decay of ΔOD extends over several hundreds of ps, it cannot solely stem from vibrational relaxation in the triplet manifold. This feature indicates a photoreaction leading to intermediates and/or stable photoproducts which reveal high absorption cross sections in a spectral range of 440 to 560 nm. Moreover, despite continuously pumping, decomposition of the samples caused by laser irradiation was observed. Photolysis of the *meta* and *para* isomers performed in static cuvettes confirms this finding. Photodegradation was found for both isomers, as exemplarily shown for *p*-NP in 2-propanol in figure B.6 in appendix B.

Additionally, it is known for several nitrated polycyclic aromatic hydrocarbons (NPAHs), that evolution of the excited singlet state occurs via two parallel pathways: i) ultrafast ISC followed by population of the first excited triplet state T_1 and ii) photochemical dissociation culminating in release of nitrogen(II)oxide (NO·) and the corresponding aryloxy radical (Ar–O·).[161-163] In the following, both relaxation channels are described more precisely.

Deactivation of the initially excited singlet state by ISC was found on fs to sub-ps time scale for most NPAHs.[162] According to the El-Sayed rules,[156, 157] this relaxation process was supported by efficient coupling between the $^1(\pi\pi^*)$ excited singlet state and upper triplet states revealing at least partial $^3(n\pi^*)$ character. In addition, several theoretical studies predicted nearly isoenergetic character of the involved states.[158, 164] Besides that, evidence of various photoproducts after excitation indicated an additional relaxation channel resulting in the formation of Ar–O· and NO· radicals which serve as precursor molecules for secondary reactions, as e.g. hydrogen abstraction or formation of quinones.[165-167] The suggested radical intermediates were detected by ESR spectroscopy[168] and laser photolysis with ps and ns resolution.[166, 167, 169] In this context, two types of mechanistic pathways have been considered for the radical generation over the years and are still under debate.

In 1966, Chapman and coworkers supposed dissociation via an intramolecular rearrangement of the nitro group to an oxaziridine-type ring and/or a nitrite intermediate to explain the photochemistry of 9-nitroanthracene in solution.[170] This mechanism is known as the

Chapman`s orientation-photoreactivity and is schematically illustrated in figure 4.18. The rationale behind this consists in a correlation of the capability of a photoreaction with an out-of-plane arrangement of the nitro group relative to the aromatic moiety as found for 9-nitroanthracene in the ground and the excited singlet state. As a result of the nearly perpendicular orientation, orbital overlap of the p-atomic half-vacant (non-bonding) orbital of the oxygen atom of the nitro group and the adjacent p-atomic orbital of the carbon atom of the aromatic ring favors formation of an oxaziridine-type intermediate. Further decomposition is suggested to generate a nitrite intermediate and finally NO· and the corresponding Ar–O· radicals. Direct evidence of the accumulation of the aryloxy radical of 9-nitroanthracene (anthryloxy radical) was firstly provided by Peon and coworkers using fs-resolved broad band absorption spectroscopy few years ago.[163] After excitation at 385 nm an additional increase of ΔOD on the ps time scale was found at a probe wavelength of 350 nm associated with the absorption of the anthryloxy radical. Thereby, the time constant of radical formation was found to be at 8.7 ps, and thus in similar time range to ISC and vibrational relaxation in the triplet manifold.[163] Furthermore, several studies supposed that photoinduced dissociation reaction of NPAHs takes place in the unrelaxed triplet manifold after ISC. Direct formation from the excited singlet state, however, could not completely be excluded.[161, 167, 169] For many years, orientation of the nitro group has been reckoned as the controlling aspect with regard to photodegradation of NPAHs, but oxaziridine or nitrite intermediates could not be identified due to their instability.[171]

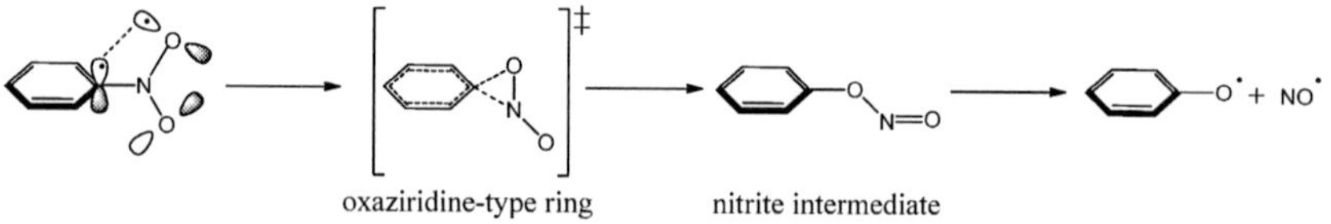

oxaziridine-type ring   nitrite intermediate

Figure 4.18: Sketch of intramolecular rearrangement of the nitro group to an oxaziridine-type ring and a nitrite intermediate supposed by Chapman et al.[170] Decomposition of the nitrite intermediate leads to release of the NO· and the corresponding Ar–O· radical.

Another mechanistic pathway for Ar–O· and NO· radical generation was proposed for 1-nitropyrene. In contrast to 9-nitroanthracene, several theoretical studies revealed a nearly planar excited state structure for 1-nitopyrene.[172, 173] But nonetheless, formation of photoproducts caused by Ar–O· radicals as precursors was found, implying that 1-nitropyrene does not follow Chapman`s relationship.[172] In this regard, direct dissociation-recombination mechanism, as schematically depicted in figure 4.19, is discussed. Presumably the C–N bond cleaves in the initially excited singlet state favored by bond stretch of approximately 0.037 Å compared to ground state resulting in the formation of aryl (Ar·) and nitrogen(IV)oxide (NO_2·) radicals.[172] Since diffusion of these primary generated radicals may be hindered by solvent environment, instantaneous geminal radical recombination is proposed. The formed nitrite intermediate subsequently dissociates into Ar–O· and NO· radicals.[172-174] The supposed primary Ar· and NO_2· radicals as well as nitrite intermediates, however, could not be identified yet.

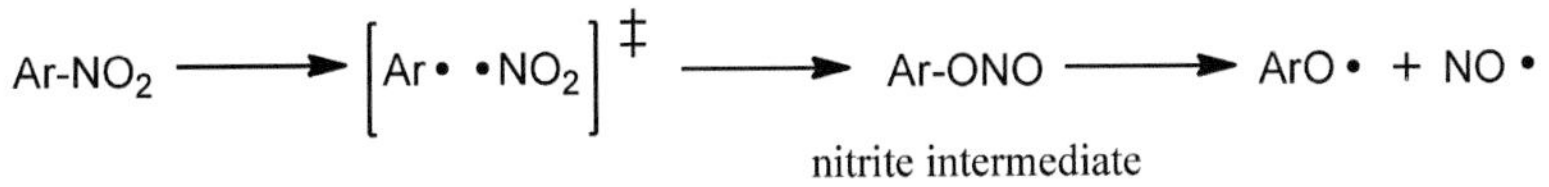

$$Ar\text{-}NO_2 \longrightarrow \left[Ar\bullet\ \bullet NO_2 \right]^{\ddagger} \longrightarrow Ar\text{-}ONO \longrightarrow ArO\bullet\ +\ NO\bullet$$

nitrite intermediate

Figure 4.19: Dissociation-recombination mechanism suggested to generate aryl and nitrogen(IV)oxide radicals in a solvent cage.[172, 173] Subsequent geminate radical recombination results in the formation of a nitrite intermediate followed by dissociation into NO· and the corresponding Ar–O· radical.

4.3.4.2 Photophysics and photochemistry of *meta*- and *para*-nitrophenol (*m*- and *p*-NP)

As has already mentioned in the section before, the transient features observed for *m*- and *p*-NP highly indicate population of triplet states as well as photochemical degradation. Since the intermediate rise of ΔOD reveals high analogy to the findings of Peon *et al.*[163] with respect to the formation of the aryloxy radical of 9-nitroanthracene, decomposition into nitrogen(II)oxide (NO·) and the corresponding aryloxy radical (Ar–O·) is proposed for *m*- and *p*-NP.

Nevertheless, the transient absorption spectra (see figure 4.14 and 4.15 a) and b)) show some differences between both isomers: ΔOD of *p*-NP reveals two spectrally distinguishable absorption bands, whereas for *m*-NP only one broad band is observable including a spectral shift to the blue within 5-10 ps after time zero. On the one hand different absorption cross sections of the suggested Ar–O· radicals as well as photoproducts may influence ΔOD dependent on probe wavelengths. Since information about the electronically excited species, however, is not accessible, this effect on TA cannot be quantified. On the other hand relaxation into upper triplet states, vibrational relaxation therein and population of T_1 must be considered as a possible parallel pathway and very likely superimpose TA features of the photoreaction. Thus, the excitation wavelengths of vertical triplet transitions can provide further insights into the relaxation behavior. DADS spectra together with calculated selected triplet excitations related to the optimized T_1 geometry (for Cartesian coordinates see appendix B table B.14 and B.15) are displayed in figure 4.20 and 4.21 for p-NP and m-NP, respectively. Considering the spectral range between 390-610 nm, one triplet transition at about 479.3 nm is found for p-NP, whereas for m-NP four triplet excitations are identified (at about 597.5, 524.2, 451.0 and 397.5 nm). The numerical values of the computed triplet transitions are listed in appendix B table B.22 and B.23 along with their corresponding oscillator strengths. Regarding observed ΔOD as composition of ESA of the triplet manifold (of the original nitrophenol molecules) and generated aryloxy radicals (Ar–O·) as well as secondary photoproducts, the absorption band around 480 nm found for p-NP is likely caused by resonant triplet absorption. The vertical transition, however, can only be considered as an indication for possible triplet absorption as they are related to the optimized T_1 state. For m-NP, several triplet transitions are found which contribute to ΔOD in agreement with the observed spectral broad absorption band (see figure 4.21 a) and b)). Thereby, the shift to the blue within few ps after time zero is very likely attributable to vibrational relaxation.[175]

In summary, the following picture of photochemical relaxation of m- and p-NP emerges in consideration of photodegradation and triplet formation. For p-NP, the fast decay after time zero associated with τ_1, and the corresponding amplitude A, is assigned to depopulation of S_2 via ISC into upper triplet states. Since this process occurs within 1.8-4.2 ps dependent on solvent, it very likely competes with IC into the ground state and/or the underlying dark singlet state S_1. Hence, τ_1 can only be reckoned as a lower limit of ISC. Moreover, orbital analysis indicates that ISC presumably takes place from the excited singlet $^1(\pi\pi^*)$ state into T_4 which reveals $^3(n\pi^*)$ character according to the El-Sayed rules[156, 157] and is also facilitated by a small energy gap of approximately 0.40 eV (related to the optimized structures).

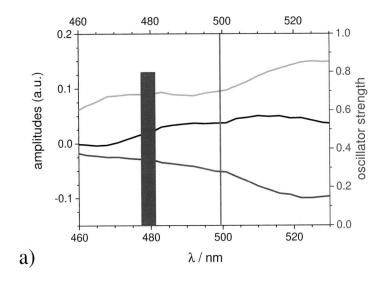

a)

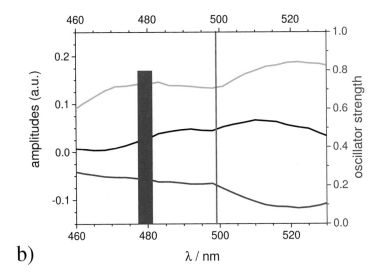

b)

Figure 4.20: DADS for *p*-NP in **chloroform** (a) and **2-propanol** (b) together with selected calculated transitions related to the optimized geometry of T_1 (blue bar). The corresponding oscillator strength is plotted on the right axis. Amplitude A is represented by black, amplitude B by red and amplitude C by green lines. The grey lines in a) and b) indicate the different spectral regions which are separately analyzed by the global fit routine.

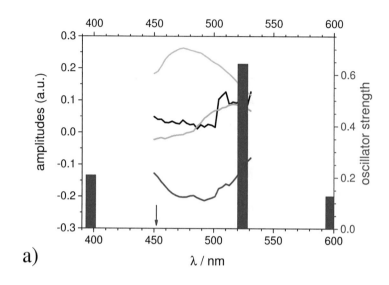

a)

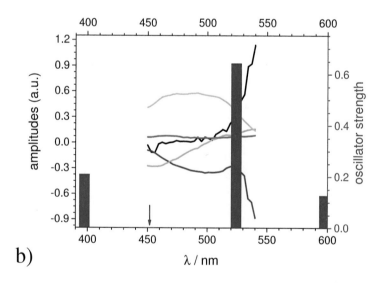

b)

Figure 4.21: DADS spectra for *m*-NP in **chloroform (a)** and **2-propanol (b)** together with selected calculated transitions related to the optimized geometry of T_1 (blue bars, blue arrow). The corresponding oscillator strengths are plotted on the right axis. Amplitude A is represented by black, amplitude B by red, amplitude C by green, amplitude D by cyan and (in case of 2-propanol) amplitude E by dark green lines.

The rise of ΔOD on the ps time scale associated with τ_2 and amplitude B is attributed to accumulation of the aryloxy radicals (Ar–O·) which agrees well with detection of the anthryloxy radical of 9-nitroanthracene conducted by Peon *et al.*[163] Consequently, the following decrease of ΔOD is associated with radical reactions resulting in τ_3 and amplitude C. Since triplet relaxation is a parallel pathway, the obtained time constants (and amplitudes) are superimposed by vibrational relaxation into the triplet manifold and depopulation of T_1. As can be seen from figure 4.20, amplitude B reveals a maximum value around 520 nm indicating a high amount of radical absorption in this spectral region. Slightly higher time constants are found for global fitting of the absorption bands around 475 nm in comparison to the band around 525 nm due to various contributions of triplet and radical absorption on TA dependent on probe wavelengths.

In analogy to *p*-NP, τ_1 and amplitude A derived for *m*-NP is related to depopulation of S_1 via ISC within ≤0.1-0.2 ps into upper triplet states competed by IC into the ground state. As orbital analysis shows, ISC most likely occurs between the $^1(\pi\pi^*)$ S_1 state and T_3 revealing $^3(n\pi^*)$ character. The energy gap (related to optimized structures) is found to be 0.36 eV. τ_2 (amplitude B) and τ_3 (amplitude C) are assigned to vibrational relaxation into the triplet manifold and formation of aryloxy radicals of *m*-NP. To model the blue shift τ_2 has to be optimized separately for each probe wavelength. The decrease of ΔOD is expressed by τ_4 (amplitude D) and is mainly related to radical reactions. Additionally, the time constants are likely imposed by vibrational triplet relaxation and depopulation of T_1 of *m*-NP. In 2-propanol, the latter process can be assigned to the observed constant long time absorption associated with τ_5 (amplitude E). Further contributions arising from absorption of generated photoproducts cannot completely be excluded. As can directly be seen from figure 4.21, amplitudes associated with radical and triplet absorption (B and D) show one maximum indicating that absorption properties of both species are overlapping over the entire wavelength range.

4.3.4.3 Role of the dissociation mechanism

The spectral evolution in *p*- and *m*-NP is highly complex due to the appearance of ESA of the two parallel relaxation channels, triplet and radical formation, on a similar time scale. Since depopulation of the initially excited singlet states is observed (associated with τ_1), radical generation most likely originates from the upper unrelaxed triplet manifold, but cannot completely be excluded from the excited singlet states. The proposed pathways of *m*- and *p*-NP in chloroform and 2-propnaol are simplified in figure 4.23.

In order to gain further insights into the mechanism of radical formation – previously discussed intermolecular nitro-nitrite-rearrangement or dissociation-recombination mechanism – orientation of the nitro group is of possible interest. Structural information about unrelaxed triplet states, however, is not easily accessible. Nonetheless, geometry optimization of the initially excited singlet states (for Cartesian coordinates see appendix B table B.8 and B.9) can provide indications of structural changes which may also affect the upper unrelaxed triplet states. The geometry optimized structures of *m*-NP (S_1) and *p*-NP (S_2) are shown in figure 4.22 a) and b). For *p*-NP, a bend of the two oxygen atoms toward the aromatic ring is found resulting in a dihedral angle (ONCC, labeled by * in the graph) of ~20°, whereas *m*-NP stays planar in its exited state. Besides that, electron distribution of LUMOs of both isomers (see figure 4.13) hints at antibonding character with respect to the C-N bond. The optimized C-N distance is found to be 1.521 Å for *p*-NP in S_2, and thus enlarged by ~0.045 Å compared to S_0 (1.476 Å). On the contrary, the C–N bond for the relaxed *m*-NP in S_1 is shortened by about 0.014 Å related to the ground state (C–N distance S_1 1.474 Å, S_0 1.488 Å). Hence, no clear proof of either nitro-nitrite-rearrangement or dissociation-recombination mechanism could be achieved by optimizing the initially excited singlet states. Furthermore, experimental evidence of suggested intermediates, e.g. oxazaridine-type ring and/or Ar· and NO_2· radicals, was not provided by the obtained TA data most likely due to the high instability and reactivity of these species. It is concluded that the question on the mechanistic process of Ar–O· and NO· radical generation could not be unambiguously and conclusively clarified at this point. Hence, further ab-initio calculations are required which maps the different channels of radical formation in singlet as well as in triplet states. Moreover, the inclusion of the solvent environment can provide more insights into the photodissociation mechanism of *m*- and *p*-NP.

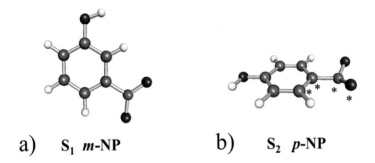

a)   S$_1$ *m*-NP   b)   S$_2$ *p*-NP

Figure 4.22: Geometry optimized structures of S$_1$ of **m-NP (a)** and S$_2$ of **p-NP (b)** at the TD-DFT level of theory. A dihedral angle (ONCC, labeled by * in the graph) of ~20° was found for *p*-NP, whereas *m*-NP stays planar in its initially excited state.

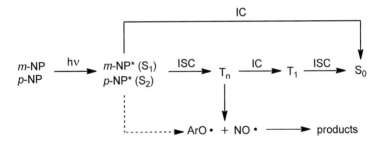

Figure 4.23: Suggested relaxation pathways of *m*- and *p*-NP in **chloroform** and **2-propanol** after excitation into S$_1$ (*m*-NP) and S$_2$ (*p*-NP). Photodegradation is supposed to result in the formation of the nitrogen(II)oxide (NO·) and the corresponding aryloxy radical (Ar–O·) which serve as precursors for secondary photoproducts.

4.3.5 Dynamics in water

In contrast to the dynamics of *p*- and *m*-NP observed in chloroform and 2-propanol, no growth of ΔOD is detectable in water on the ps time scale (see figure 4.14 and 4.15 c)). Thus, photoreaction likely plays either a minor role or does not occur in this case. Considering the aforementioned parallel relaxation channel via ultrafast ISC after excitation, population of triplet states is suggested to be the main pathway in water.

For *p*-NP, the DADS spectrum is displayed in figure 4.24 a). The remaining weak TA on the hundreds of ps time scale associated with τ_3 (amplitude C) is assigned to depopulation of T_1 being in line with the calculated triplet transition at about 479.3 nm. Consequently, the second time constant τ_2 (amplitude B) is attributable to vibrational relaxation in the triplet manifold manifested in the observed spectral shift to the blue in the white light generated TA spectrum (see figure 4.14 c)). However, since no dependence of τ_2 on probe wavelengths was found (see section 4.3.2.2), contribution of the spectral shift to ΔOD seems to be very low. Furthermore, amplitude B reveals two maxima (around 477 and 495 nm) which correspond to the appearance of the observed distinguishable absorption bands (centered at 475 and 507 nm) in the TA (wl) spectrum. Obviously, the occurrence of the red band is caused by formation of aryloxy radicals. Its influence on the overall TA, however, is too weak to be detected in the obtained time constants (see section 4.3.2.2). Finally, τ_1 (amplitude A) is assigned to ISC into upper triplet states likely superimposed by IC into the ground state.

Since, for *m*-NP, ΔOD returns to zero within several ps, no hints of a long-lived species are given. This behavior indicates strong quenching of excited states by the solvent environment. The DADS spectrum of *m*-NP in water is shown in figure 4.24 b). In analogy to *p*-NP, τ_1 (amplitude A) is related to depopulation of S_1 via ISC into upper triplet states and IC into the ground state, whereas τ_2 (amplitude B) is corresponding to vibrational relaxation into the triplet manifold. Due to the absence of any kind of radical formation it is speculated that photodegradation of *m*-NP is completely suppressed by efficient coupling and energy transfer to solvent molecules. For this reason, the observations in section 4.3.2.1 and section 4.3.2.2 regarding the solvent dependence of TA are explainable: Decrease of intensity of the absorption band centered at 525 nm and increase of the band around 480 nm is found for *p*-NP in the order of chloroform, 2-propanol and water, and thus, from non-polar to polar solvents (for dipole moments see table 4.1). This behavior implies that relaxation via ISC seems to become a more dominant relaxation channel in polar solvents.

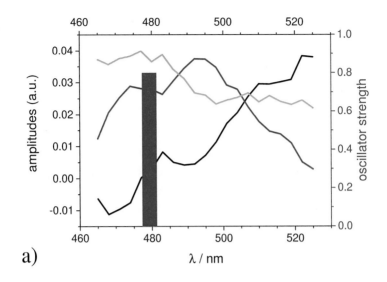

a)

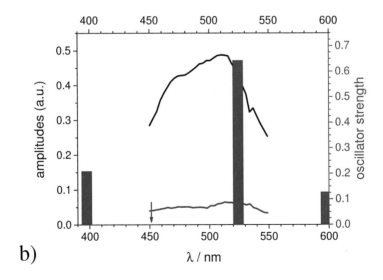

b)

Figure 4.24: DADS spectra for *p*-NP (a) and *m*-NP (b) in **water** together with selected calculated transitions related to the optimized geometry of T_1 (blue bars, blue arrow). The corresponding oscillator strengths are plotted on the right axis. Amplitude A is represented by black, amplitude B by red and (in case of *p*-NP) amplitude C by green lines.

For *m*-NP, faster dynamics are found in the order of chloroform, 2-propanol and water. Both facts point to quenching processes of excited states which are facilitated in polar solvents and finally lead to suppression of photodegradation after excitation into the first bright singlet state. Nevertheless, one still open issue is the origin of the observed SE for *p*- and *m*-NP in water. In contrast to *o*-NP, geometry optimizations of the initially excited singlet states of the isomers provide no indications of high structural changes and a small energy gap between the excited and the ground state. Since NPs are weak acids (for pK_a-values see figure 4.1) the question arises whether formation of phenolates may play a role on the excited state dynamics in water. Following this consideration, there are two possibilities of phenolate generation: i) equilibrium formation between protonated and deprotonated forms and ii) formation from excited states due to intermolecular proton transfer to solvent molecules. Considering the instantaneous appearance of SE after time zero (figure 4.15 c) and 4.17), the latter process is supposed to take place in the excited singlet state within few hundreds of fs. The proposed relaxation pathways in water are summarized in figure 4.25. In order to provide clarity, deprotonated forms of *m*-, *p*-NP and also *o*-NP are investigated with both steady state UV-Vis and time-resolved pump-probe absorption spectroscopy. The following sections contain absorption and excited state properties of the nitrophenolates as well as further analyses by means of quantum chemical calculations. A complete overview of the highly complex dynamics of nitrophenols and the interplay of protonated and deprotonated forms is provided at the end of this chapter.

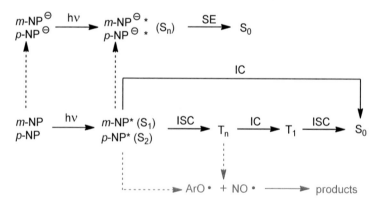

Figure 4.25: Suggested relaxation pathways of *m*- and *p*-NP in **water**. Photoreaction likely plays either only a minor role or does not occur in water. Generation of nitrophenolates is suggested to cause ultrafast SE.

4.4 Dynamics of nitrophenolates after intramolecular charge transfer excitation

4.4.1 Steady state spectroscopy and calculated vertical transitions

To investigate the potential influence of solvent polarity, of hydrogen bonding, as well as of different counter ions on steady state and time-resolved spectra, two ways of nitrophenolate formation were chosen: Deprotonation was performed by i) sodium hydroxide (NaOH) in water and ii) 1,8-diazabicyclo[5.4.0]undec-7-ene (DBU) in chloroform (see section 3.5.1). Figure 4.26 shows direct comparison of UV-Vis spectra of protonated and deprotonated forms.

A red-shift of approximately 60-90 nm depending on the isomer is observed for all nitrophenolates. Furthermore, spectra of deprotonated forms are characterized by absorption minima at the lowest-energy absorption maxima of their corresponding protonated forms. For p-NP$^{\ominus}$, the first absorption band is centered at 400 nm (chloroform) and 401 nm (water) which agrees well with previously recorded spectra in different solvents using sodium methoxide as base.[176] As one can see, the maximum wavelengths of the observed transitions are nearly independent of solvent and employed counter ions, whereas slightly higher influence of environment is found for m-NP$^{\ominus}$ and o-NP$^{\ominus}$. In this case, the lowest-energy absorption maxima are centered at 395 nm (m-NP$^{\ominus}$) and 428 nm (o-NP$^{\ominus}$) in chloroform as well as at 390 nm (m-NP$^{\ominus}$) and 416 nm (o-NP$^{\ominus}$) in water, respectively. In general, a decrease of the extinction coefficients is discovered in the order of the *para*, *ortho* and *meta* form. Values obtained at the maximum wavelength (λ_{max}) of the first absorption band are listed in table 4.8 and are in line with a previous study in acetonitrile and in the gas phase.[160] The extremely low oscillator strengths of m-NP$^{\ominus}$ (factor 3-14 smaller compared to its isomers) are associated with its stronger CT character caused by weaker coupling between donor and acceptor moieties in contrary to p-NP$^{\ominus}$ and o-NP$^{\ominus}$.[160] Even though one would expect the smallest excitation energy for m-NP$^{\ominus}$ for that reason, the first absorption band is blue-shifted in comparison with the *ortho* and *para* form in solution. However, as known from experimental and theoretical studies of m-NP$^{\ominus}$,[145, 160] solvent interaction lowers the ground state energy much stronger than the excited state energy (favored by the charge distribution of the *meta* form in S_0) which results in the observed high excitation energies in solution.

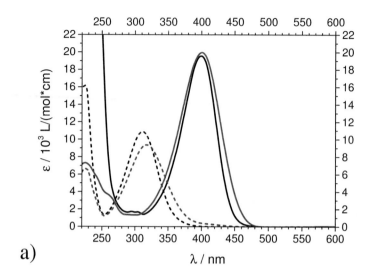

a)

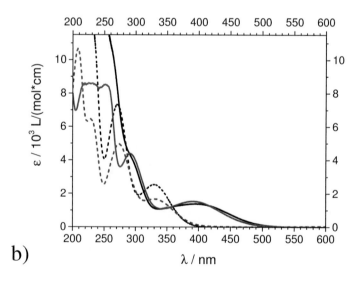

b)

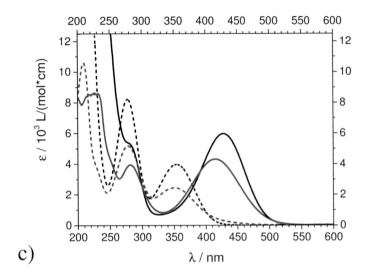

c)

Figure 4.26: Steady-state spectra of **p-NP/p-NP$^\ominus$** (a), **m-NP/m-NP$^\ominus$** (b) and **o-NP/o-NP$^\ominus$** (c) in chloroform (black) and water (red). Protonated forms are represented by dashed and deprotonated forms by solid lines.

Table 4.8: Extinction coefficients obtained at the maximum wavelength (λ_{max}) of the first absorption bands for p-NP$^\ominus$, m-NP$^\ominus$ and o-NP$^\ominus$ in different solvents.

nitrophenolate		chloroform + DBU	water + NaOH
p-NP$^\ominus$	λ_{max}/nm	400	401
	$\varepsilon/10^3$L/(mol·cm)	19.55	19.97
m-NP$^\ominus$	λ_{max}/nm	395	390
	$\varepsilon/10^3$L/(mol·cm)	1.38	1.53
o-NP$^\ominus$	λ_{max}/nm	428	416
	$\varepsilon/10^3$L/(mol·cm)	5.97	4.30

In analogy to the nitrophenols, all deprotonated isomers exhibit a planar minimum geometry in their electronic ground state corresponding to C_S symmetry (see appendix B table B.4–B.6 for the computed coordinates of the optimized ground state structures). Selected vertical singlet-singlet transitions with A' and A" symmetry as well as singlet-triplet transitions at the Franck-Condon region are displayed in figure 4.27. The calculations were performed at the TD-DFT level of theory (see section 3.6) and related to the ground states 1A' (S_0). As for the nitrophenols, the oscillator strengths obtained for singlet-singlet transitions with A" symmetry are smaller by several orders of magnitude. Hence these excitations are related to transitions into dark singlet states. The computed values agree well with the recorded spectra in solution, however, calculated singlet-singlet transitions have to be shifted to the blue by about 1 eV to match experimental data of m-NP$^\ominus$. For the sake of clarity, only selected singlet states with A' symmetry are explicitly labeled in figure 4.27. The employed excitation wavelength of the time-resolved experiments was chosen to be at 418 nm for all isomers in order to provide sufficient excitation energy (see section 3.5.1). Considering the calculations, excitation at 418 nm corresponds to a transition into the first excited bright singlet state for all nitrophenolates. Thereby, this transition is attributable to the second excited singlet state in the case of p-NP$^\ominus$ and o-NP$^\ominus$ (2A' (S_2)), whereas the first excited singlet state (2A' (S_1)) is included in the case of m-NP$^\ominus$.

As inspection of the molecular orbitals indicates, excitation into the first excited bright singlet state is associated with intramolecular charge transfer from the phenolate oxygen atom to the nitro group corresponding to a transition into a $^1(\pi\pi^*)$ state for all isomers. These findings are in line with previous calculations.[160] The involved orbitals are displayed in figure 4.28 (a) p-NP$^\ominus$, b) m-NP$^\ominus$, c) o-NP$^\ominus$). For p-NP$^\ominus$ and m-NP$^\ominus$, excitation into the first bright singlet state is assigned to a transition from the HOMO into the LUMO, whereas for o-NP$^\ominus$ the LUMO-1 is involved instead of the LUMO. Analogously to the protonated forms, the HOMO orbitals of the phenolates are characterized by a π-MO revealing electron distribution delocalized over the phenolate oxygen atom and the benzene ring and, in the case of p-NP$^\ominus$ and o-NP$^\ominus$, also over the nitro group. The LUMOs as well as the LUMO-1 are described by π^*-MO mainly implying the benzene ring and the nitro group. As mentioned above, only a slight dependence on solvent environment and counter ions are found which is attributable to different extents of ground and excited states stabilization. Furthermore, TD-DFT calculations exhibit partly nearly isoenergetic triplet states below the first bright excited singlet states. For simplification, symmetry notation of the excited states is neglected in the following sections.

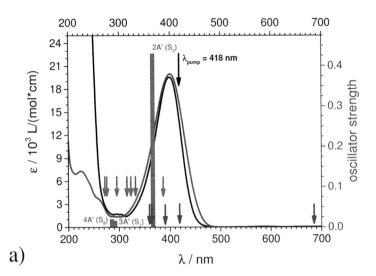

a)

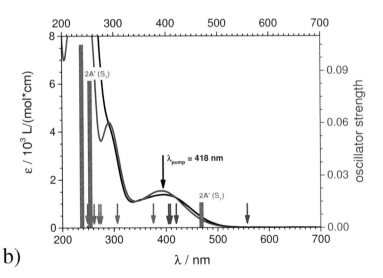

b)

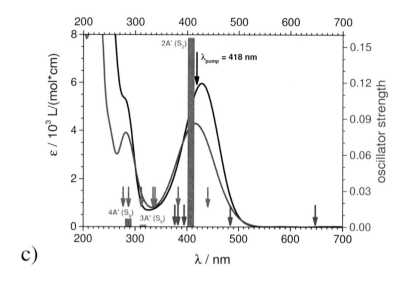

c)

Figure 4.27: Absorption spectra of *p*-NP$^\ominus$ **(a)**, *m*-NP$^\ominus$ **(b)** and *o*-NP$^\ominus$ **(c)** in chloroform (black) and water (red). Calculated oscillator strengths of selected singlet-singlet transitions with A' symmetry are displayed as grey bars (right axis). Singlet-singlet transitions with A" symmetry are marked with grey and singlet-triplet transitions with blue arrows. Singlet-singlet transitions of *m*-NP$^\ominus$ are shifted to the blue by about 1 eV. The excitation wavelength of time-resolved experiments was at 418 nm for all isomers.

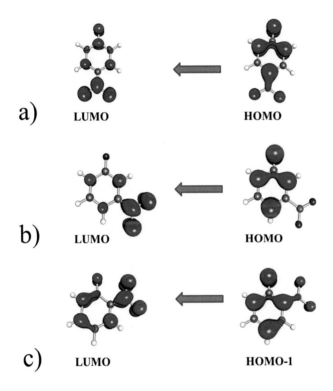

Figure 4.28: Involved orbitals in the charge transfer transition into the first excited bright singlet state. For *p*-NP$^{\ominus}$ (a) and *m*-NP$^{\ominus}$ (b), excitation is attributable to a LUMO–HOMO transition, whereas for *o*-NP$^{\ominus}$ (c) the HOMO-1 is involved instead of the LUMO.

4.4.2 Transient absorption spectroscopy in solution

After ICT excitation into S_2 (p-NP$^{\ominus}$, o-NP$^{\ominus}$) and S_1 (m-NP$^{\ominus}$) at 418 nm, the temporal evolution of the excited state depopulation is monitored by selected probe wavelengths between 500 and 950 nm. Since TA features reveal slight solvent dependence (except for m-NP$^{\ominus}$) and partly low signal-to-noise ratio in water, only TA profiles recorded in chloroform are shown in figure 4.29. For traces performed in water see appendix B figure B.7.

As one can see from figure 4.29 a), ΔOD of p-NP$^{\ominus}$ is characterized by a superposition of SE and ESA at probe wavelengths between 500 and 700 nm. After SE within few hundreds of fs, TA returns to positive values and decays on the sub-ps time scale. Thereby, the intensity of ΔOD decreases for higher probe wavelengths and becomes nearly zero at 900 and 950 nm.

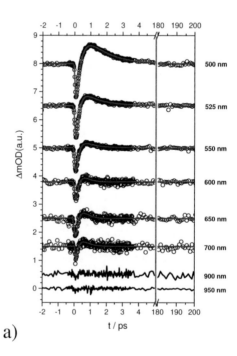

a)

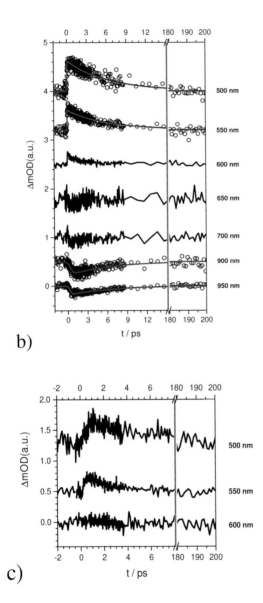

Figure 4.29: TA profiles of *p*-NP$^\ominus$ (a), *m*-NP$^\ominus$ (b) and *o*-NP$^\ominus$ (c) in **chloroform** at a pump wavelength of 418 nm and probe wavelengths as labeled. Fit functions (if possible) are displayed as red lines. TA profiles of *o*-NP$^\ominus$ could not be analyzed. Note that the plots are shifted along the ordinate for better clarity.

Similarly, ESA and SE are observed for m-NP$^\ominus$ within 10 ps (figure 4.29 b)). ΔOD reveals positive values in a probe wavelength range of 500-550 nm, whereas negative values are found at 900 and 950 nm. Strong decrease in intensity of ΔOD occurs between 600 and 700 nm resulting in small ΔOD values most likely due to a superposition of ESA and SE. Considering figure 4.29 c), ESA of o-NP$^\ominus$ shows a slow rise at time zero at 500 and 550 nm extending over one ps indicating a hidden negative component presumably caused by SE. Direct indication of SE, however, is not provided by TA measurements in chloroform. At probe wavelengths beyond 550 nm up to the NIR, ΔOD is zero as shown by the example of probing at 600 nm in figure 4.29 c). Nonetheless, weak SE is found for o-NP$^\ominus$ at 650 up to 950 nm in water within few hundreds of fs (see appendix B figure B.7 c)). In summary, it can be stated that TA features of all nitrophenolates reveal high analogy to those observed for o-NP (see section 4.2). In contrast to o-NP, however, no residual absorption is detected for the deprotonated forms with respect to long delay times.

For a quantitative investigation of the observed features, the data are analyzed by fit functions described in section 3.4. The obtained time constants in chloroform as well as water are summarized in table 4.9. In contrast to o-NP, ESA and SE could not be analyzed separately because they overlap in the entire range of employed probe wavelengths. Global analysis between 500 and 700 nm achieves best results with a sum of three exponentials. In this regard, the time constant associated with SE is named τ_{ind} in analogy to the denotation used for o-NP. Further time constants (τ_1 and τ_2) are attributable to an ultrafast decay of ESA and additional dynamics on a sub-ps time scale. Likewise three exponentials are required to model TA of p-NP$^\ominus$ in water, whereby τ_{ind} was extracted from SE signals in the NIR which are not superimposed by ESA. As can be seen from table 4.9, equal values for τ_{ind} and τ_1 are determined in both solvents. For m-NP$^\ominus$, monoexponential decay was found for ESA at 500 and 550 nm as well as SE in the NIR (900 and 950 nm). Since identical time constants were obtained TA could be analyzed by a global fit routine. These findings indicate that ESA does not consist of contributions of several excited species. To comply with the introduced notation of time constants for o-NP the obtained value is listed twice: as τ_{ind} (SE) and τ_1 (ESA). TA profiles recorded at probe wavelengths in between are not included in the evaluation due to low intensity and poor signal-to-noise ratio. In the same way, monoexponential decay is found for m-NP$^\ominus$ in water at 500 and 900 nm resulting in a significant faster time constant. Obtained fit functions of p-NP$^\ominus$ and m-NP$^\ominus$ in chloroform are displayed in figure 4.29 a) and b) as red lines together with the corresponding TA traces. Fit functions obtained in water are shown in appendix B figure B.7. In addition, quantitative investigation of ΔOD of o-NP$^\ominus$ was neither possible in chloroform nor in water due to low intensity and poor signal-to-noise ratio.

Comparing the results from table 4.9, SE takes place on a similar time scale as observed for o-NP (0.2–0.3 ps dependent on solvent, section 4.2.2, table 4.3). One exception is m-NP$^\ominus$ which shows high dependence of the time constants on the employed solvent. As a result, the value obtained in chloroform is approximately a factor of 21 higher than in water.

Table 4.9: Time constants of p-NP$^\ominus$ and m-NP$^\ominus$ obtained by global fit routine in chloroform and water. The given errors are standard deviations derived from the fitting procedure. TA traces of o-NP$^\ominus$ could not be analyzed. For more details, see text above.

		p-NP$^\ominus$		m-NP$^\ominus$	
		chloroform	water	chloroform	water
SE	τ_{ind}/ps	0.3 ± 0.2	0.2 ± 0.1	6.2 ± 0.2	0.3 ± 0.1
ESA	τ_1/ps	0.3 ± 0.2	0.2 ± 0.1	6.2 ± 0.2	0.3 ± 0.1
	τ_2/ps	1.3 ± 0.1	0.6 ± 0.1	---	---

4.4.3 Quantum chemical calculations

4.4.3.1 Density functional theory

Taking into account the instantaneous SE of o-NP related to structural changes in S_1 (ESIPT accompanied by out-of-plane rotation of the HONO group, section 4.2.3), similar excited state properties are predicted for the nitrophenolates. Hence, geometry optimization of the initially excited singlet states at the TD-DFT level of theory can provide first indications of possible molecular changes. For Cartesian coordinates see appendix B table B.10–B.12.

Optimization of the S_2 state of p-NP$^\ominus$, however, yields a planar minimum structure as also found for its electronic ground state. An analog attempt to optimize the geometry of the dark singlet state (S_1) lying below S_2 failed. Here, the calculation points to energy stabilization due to an out-of-plane rotation of the NO_2 group. Nonetheless, a stable energy minimum could not be identified. The computed structure of S_1 is shown in figure 4.30 a) and characterized by a dihedral angle (ONCC, labeled by * in the graph) of ~65°. Since SE is observable in a wide probe wavelength range (from the visible up to the NIR), it is assumed that relaxation into S_1 plays an important role in the photophysical pathways after excitation of p-NP$^\ominus$ at 418 nm. In contrast to p-NP$^\ominus$, a decrease in the S_2 energy of o-NP$^\ominus$ (initially excited singlet state) is found due to torsion of the NO_2 group resulting in a dihedral angle ONCC of ~102° (see figure 4.30 b)). Likewise, no stable minimum structure could be computed. Moreover, the dark singlet state (S_1) of o-NP$^\ominus$ reveals a planar minimum geometry. Finally, no evidence of an out-of-plane rotation of the NO_2 group is found by optimizing the initially excited singlet state S_1 of m-NP$^\ominus$. The optimized structure (figure 4.30 c)) is nearly planar with a dihedral angle ONCC slightly different from zero (0.3°). As mentioned previously, no dark singlet states lie below the first bright state. Despite of this, SE was observed within 10 ps in TA experiments (figure 4.29 b)) whose origin needs to be clarified.

In summary, it can be stated that excited state calculations of p-NP$^\ominus$ and o-NP$^\ominus$ agree well with the results obtained for o-NP at the TD-DFT level of theory (see section 4.2.3.1). To get more detailed insights into the relative energetic positions of the excited and the ground state PES, and to answer the question why the geometry optimization in the case of m-NP$^\ominus$ does not provide any indications for a twist of the NO_2 group, further calculations using multireference methods were performed.

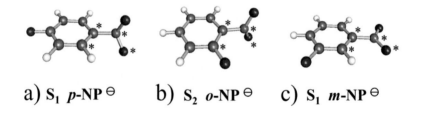

a) S_1 p-NP $^\ominus$ b) S_2 o-NP $^\ominus$ c) S_1 m-NP $^\ominus$

Figure 4.30: Optimized structures of p-NP$^\ominus$ (a), o-NP$^\ominus$ (b) and m-NP$^\ominus$ (c) for excited singlet states as labeled at the TD-DFT level of theory. Dihedral angles ONCC are marked by * in the structures and calculated to be ~65°, ~102° and ~0.3° in the order of the *para*, *ortho* and *meta* form.

4.4.3.2 Multireference methods

Calculations at the CASSCF level of theory were performed in collaboration with C. García Iriepa (PhD student, Universidad de Alcalá, Spain) during a short stay at the Karlsruhe Institute of Technology (see section 3.6).

Figure 4.31 displays the minimum energy pathway starting at the Franck-Condon region in the initially excited bright state of p-NP$^\ominus$ (S$_2$) using a complete active space which consists of 10 electrons and 8 orbitals ((10, 8)-CAS). In doing so, the relative energy (E$_{rel}$, in eV) is obtained as a function of the C–N stretching coordinate (b$_1$) and the ONCC torsion angle (φ) as defined in the molecular structure inserted in figure 4.31.

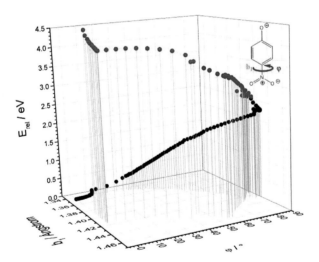

Figure 4.31: Minimum energy pathway of p-NP$^\ominus$ starting at the Franck-Condon region. The relative energy (E$_{rel}$) of S$_2$ (red points), S$_1$ (blue points) and S$_0$ (black points) is plotted against the C–N stretching coordinate (b$_1$) and ONCC torsion angle (φ). The definition of b$_1$ and φ is shown with the help of the molecular structure inserted.

Upon excitation into S_2, a flat region on the PES is reached, characterized by a relaxed planar structure due to activation of the C–N stretching mode. Thereby, b_1 is enlarged by 0.027 Å compared to the ground state (b_1 (S_0) = 1.388 Å; b_1 (S_1) = 1.415 Å). These findings are in agreement with the aforementioned TD-DFT calculations predicting a planar minimum structure of S_2. At that point, torsion around the C–N bond (angle φ) hardly leads to a change of the energy up to a torsion angle of approximately 50°. Then, CI with the first excited singlet state S_1 is reached. Subsequently, a high decrease in energy evolves in S_1 by further increase of φ leading to a close approach of the S_1 and S_0 PES. Finally, at a torsion angle of ~90°, CI with S_0 is found in line with previous TD-DFT calculations (see section 4.4.3.1). As shown in figure 4.32, return into the ground state, and hence to a planar geometry, can occur due to clock- or counterclockwise rotation of the NO_2 group preceded by shortening of one of the two N–O bonds about 0.064 Å. Thereby, the calculated branch on the right side of figure 4.32 is equivalent to E_{rel} of S_0 shown in figure 4.31.

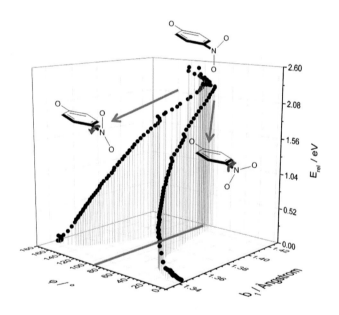

Figure 4.32: Recovery of the ground state can occur by clock- or counterclockwise rotation of the NO_2 group caused by the symmetry of ***p*-NP**$^{\ominus}$. The relative energy (E_{rel}) is plotted as a function of the C–N stretching coordinate (b_1) and ONCC torsion angle (φ) as defined in figure 4.31.

For m-NP$^\ominus$, CASSCF calculations ((10, 8)-CAS) reveal a similar behavior. The computed course of the relative energy dependent on the distance of the C–N bond (b_1) and the torsion angle φ is displayed in figure 4.33. After excitation into S_1, m-NP$^\ominus$ also relaxes via a stretching mode along the C–N bond reaching a planar geometry in the excited state. In contrast to p-NP$^\ominus$, however, shortening of b_1 was found to be at 0.022 Å compared to the ground state (b_1 (S_0) = 1.405 Å; b_1 (S_1) = 1.383 Å). Thereafter, rotation of the NO$_2$ group by ~70° leads to population of S_0 via CI. As can be seen from figure 4.33, a weak energy barrier (~21 kJ·mol^{-1}) was found along the torsion coordinate before CI with S_0 is reached. This fact is very likely due to reducing of b_1 which may hinder rotation of the NO$_2$ group around the C–N bond and explain the planar minimum structure obtained by geometry optimization at the TD-DFT level of theory. For o-NP$^\ominus$, CASSCF calculations do not provide clear indications with respect to out-of-plane rotation of the NO$_2$ group, neither for S_2 nor for S_1.

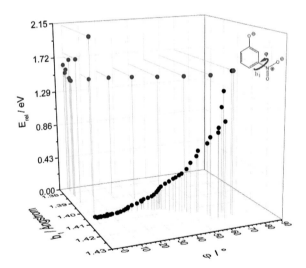

Figure 4.33: Minimum energy pathway of m-NP$^\ominus$ starting at the Franck-Condon region. The relative energy (E_{rel}) of S_1 (blue points) and S_0 (black points) is plotted against the C–N stretching coordinate (b_1) and ONCC torsion angle (φ) as defined in the inserted structure. Torsion around the C-N bond leads to a small increase in energy of ~21 kJ·mol^{-1} before CI with the ground state is reached.

4.4.4 Relaxation pathways of nitrophenolates

Considering the experimental findings in solution as well as the results of the ab initio calculations, high analogy to the relaxation features observed for o-NP is found. The CASSCF calculations confirm a similar origin of ultrafast appearance of SE for p-NP$^\ominus$ and m-NP$^\ominus$ due to an approach of the excited to the ground state PES. In contrast to o-NP, however, the drop of the energy gap is solely caused by out-of-plane rotation of the NO_2 group.

After excitation into S_2, torsion occurs in the second and first excited singlet state in the case of p-NP$^\ominus$. Since large changes of the excited state energy are only computed in S_1 and monoexponential decay of SE is observed, the corresponding time constant τ_{ind} is mainly attributed to depopulation of S_1. Therefore, relaxation via IC from S_2 into S_1 is likely too fast to be detected. Furthermore, τ_1 can be assigned to depopulation of S_1 monitored by ESA. As mentioned in section 4.4.2, a second time constant (τ_2) is required to fit TA traces of p-NP$^\ominus$ which is presumably related to hot ground state absorption (HGSA). Moreover, taking into account the relaxation pathways of o-NP (see section 4.2.4.2), ISC into upper triplet states, relaxation into the triplet manifold and population of T_1 are competitive processes. For p-NP$^\ominus$, however, no experimental evidence of long-lived excited species is given as one would expect in that case. Nonetheless, since triplet absorption is found to be highly solvent dependent and mainly observed at probe wavelengths below 500 nm for o-NP, it might be outside of the employed probe wavelength range for p-NP$^\ominus$. Thus, relaxation via ISC cannot completely be excluded.

In the same way, τ_{ind} and τ_1 are associated with depopulation of S_1 (initially excited state) of m-NP$^\ominus$ induced by the probe pulse and monitored by ESA, respectively. In contrast to p-NP$^\ominus$, the decay of TA profiles around 500 nm is monoexponential indicating that no additional relaxation processes underlie the depopulation of S_1 via IC. As has already been pointed out in the preceding discussion of o-NP (see section 4.2.4.1), τ_{ind} is not necessarily identical to a time constant describing depopulation of S_1 via CI with the ground state. Consequently, smaller values of τ_{ind} may be determined than expected for the IC process. For p-NP$^\ominus$ and m-NP$^\ominus$, however, identical values are obtained for τ_1 and τ_{ind}, and thus it is speculated that the aforementioned aspects only play a minor role. Moreover, the TA data of the phenolates do not provide an unambiguous dependence of the SE on the employed probe wavelengths for the same reasons as have already been discussed for o-NP in section 4.2.4.1. In summary, ultrafast depopulation of S_1 occurs within 0.2-0.3 ps for p-NP$^\ominus$ dependent on solvent, whereas two different time constants are found for m-NP$^\ominus$, namely 0.3 ps in water and 6.2 ps in chloroform. In the latter case, the slow decay of TA is well explainable by the slight energy barrier which is found on PES of S_1 toward CI with the ground state. Following this consideration, the faster time constant obtained in water is very likely caused by quenching processes due to solvent interactions (higher dipole moment and capability of hydrogen bonding compared to chloroform). The proposed relaxation pathways of p-NP$^\ominus$ and m-NP$^\ominus$ are summarized in figure 4.34.

For o-NP$^\ominus$, an analog relaxation behavior after excitation into S_2 is assumed because weak SE is detected in water (appendix B figure B.7). However, since no time constants could be extracted from the time-resolved data and out-of-plane rotation of the NO$_2$ group is only predicted by the geometry optimization of S_2 at the TD-DFT level of theory, a detailed evaluation of the relaxation pathways was not possible. For a better understanding of the relaxation pathways of o-NP$^\ominus$, time-resolved experiments at an additional pump wavelength can provide a higher intensity of ΔOD, and thus can allow for a more detailed investigation.

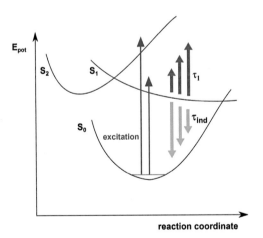

Figure 4.34: Proposed relaxation pathways of **p-NP$^\ominus$** and **m-NP$^\ominus$**. Excitation at 418 nm corresponds to transition into S_2 (p-NP$^\ominus$) and S_1 (m-NP$^\ominus$), respectively. Out-of-plane torsion of the NO$_2$ group (reaction coordinate) leads to a close approach of S_1 and S_0 PESs. Depopulation of S_1 is monitored by ESA (red arrows, τ_1) and SE (green arrows, τ_{ind}). A further time constant (τ_2) was found for p-NP$^\ominus$ which is attributed to HGSA and/or triplet relaxation and not explicitly labeled in the graph. After excitation into S_2, an analog mechanism is supposed for **o-NP$^\ominus$** which could, however, not be further analyzed.

4.5 Influence of isomerism, solvent environment and deprotonation

Reviewing the observed dynamical features of NPs and their corresponding phenolates, a highly complex relaxation behavior is found dependent on isomerism, solvent environment and deprotonation. This sections aims at the summarization of the obtained results highlighting differences as well as similarities.

Common to all investigated compounds, protonated and deprotonated, is the ICT transition from the hydroxyl group or the oxygen atom, respectively, to the nitro group upon excitation into the first bright excited singlet state. Moreover, for the protonated isomers *o*-, *m*- and *p*-NP, ultrafast ISC into upper triplet states has been identified as a characteristic relaxation pathway occurring on the sub-ps time scale. Thereby, ISC is facilitated by the El-Sayed selection rules[156, 157] as well as the nearly isoenergetic triplet states below the initially excited singlet state. The further relaxation behavior of the NPs, however, strongly depends on their different assignment of the functional groups. For *o*-NP, ESIPT combined with out-of-plane rotation of the newly formed HONO group results in minimizing the energy gap between S_1 and S_0 finally leading to a CI between these both states. The ESIPT itself is most likely triggered by the ICT character of S_1 and favored by the capability of intramolecular hydrogen bonding of the *ortho* isomer. Relaxation along this reaction coordinate was monitored by instantaneous SE induced by the probe pulses and found to be nearly independent of the employed solvents, and thus indicating a barrierless pathway.

On the contrary, high dependence on the used solvents was found for the dynamics of *m*- and *p*-NP: Photodegradation and generation of nitrogen(II)oxide (NO·) and the corresponding aryloxy radical (Ar–O·) is mainly observed in chloroform and 2-propanol, whereas ISC and population of the first excited triplet state T_1 is very likely the main relaxation channel in water. For the latter, it is further suggested that formation of the deprotonated forms leads to the observed ultrafast SE on the sub-ps time scale (for *m*-NP at 323 and 330 nm excitation; for *p*-NP at 323 nm excitation). This assumption is confirmed by investigating the dynamical features of the nitrophenolates which reveal a high analogy to those detected for *o*-NP. For the deprotonated forms, strong decrease in the excited state energy is found which is directly monitored by SE in the visible up to the NIR region. However, changes of the excited state energy are solely caused by out-of-plane torsion of the NO_2 group in contrast to *o*-NP. Following the considerations above, TA traces of *m*- and *p*-NP in water have to be reckoned as superposition of TA of protonated and deprotonated forms due to formation and excitation of both species. Since the first absorption band of all nitrophenolates is red-shifted in comparison to the corresponding phenols, SE most likely could not be observed for *p*-NP in water at an excitation wavelength of 314 nm caused by too low extinction coefficients (see section 4.4.1). As has already been mentioned in section 4.3.5, phenolate generation due to intermolecular excited state proton transfer reactions does not seem to be likely with regard to the fast appearance of SE. But nevertheless, in some cases time constants of intermolecular proton transfer were found lying in the same order of magnitude as determined for ESIPT reactions.[177] Hence, deprotonation in this way cannot be completely excluded.

In the present work, the influence of constitutional isomerism on the excited state properties of NPs and their phenolates after ICT excitation has been demonstrated. Since for *o*-NP no indications of photochemical formation of Ar–O· and NO· radicals are given, the "reactivity" of the NO_2 group with respect to this photodegradation is presumably changed by the strong intramolecular hydrogen bond. As discussed in section 4.2.5, the structural changes in *o*-NP form a cornerstone for HONO split-off. Figure 4.35 provides a summarization of the proposed relaxation pathways.

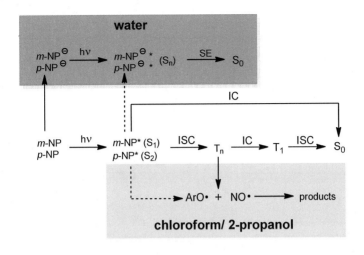

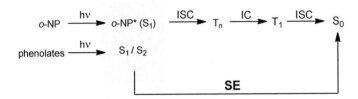

Figure 4.35: Summary of the proposed relaxation channels for **nitrophenols** and for their corresponding **phenolates** after ICT excitation into their first bright excited singlet state. Photodegradation and radical generation were found to be highly dependent on solvent for m- and p-NP, whereas instantaneous SE was detected for o-NP as well as for the nitrophenolates. For more details, see text above.

5 Ultrafast dynamics of selected type-I photoinitiators

5.1 Introduction

The work of this chapter has been conducted in collaboration with the group of Prof. Dr. C. Barner-Kowollik from the Institute for Chemical Technology and Polymer Chemistry at the Karlsruhe Institute of Technology, and has been published in part.[178] The molecules presented herein are characterized by similar ICT excitations in the UV finally leading to comparable dynamics.

Photoinduced polymerization reactions have been the subject of numerous investigations over the years.[179] In this context, apart from the monomer to be polymerized, molecules are needed which generate initiator radicals after photon absorption. Thereby, photoinitiators based on different molecular frameworks have been investigated including substituted thioxanthones[180-183], disulfides[184] and Barton esters.[185, 186] Generally, two main kinds of photoinitiators are known: i) type-I initiators which decompose into radical fragments caused by α-cleavage in the first excited triplet state[187] and ii) type-II initiators which are characterized by radical formation after abstraction of a hydrogen atom.[188, 189] In particular type-I initiators are of interest due to their large varieties of applications e.g. dental restorative materials,[190, 191] fabrication of 3-dimensional objects[192] and lithography.[89] In order to control the polymerization, and thus the chemical nature of the final polymer, it is essential to understand the ongoing processes not only with respect to the polymerization itself but also regarding the photoreaction of the initiator molecules. Hence, one has to unravel the decisive factors for an efficient polymerization reaction. In doing so, the so-called initiation efficiencies provide the information about the incorporation probabilities of different radical fragments into the polymer chain and are a measure of the ability of starting the polymerization.[91] Initiation efficiency can be determined by a post-mortem analysis of the polymerization products using pulsed laser polymerization and subsequent ionization mass spectrometry (PLP-ESI-MS).[91,][193, 194] Another crucial aspect is the number of generated radicals after excitation of the photoinitiator molecules. Thus, the fate of the initiator molecules after irradiation has to be investigated, which can be done by steady state and time-resolved spectroscopic methods.

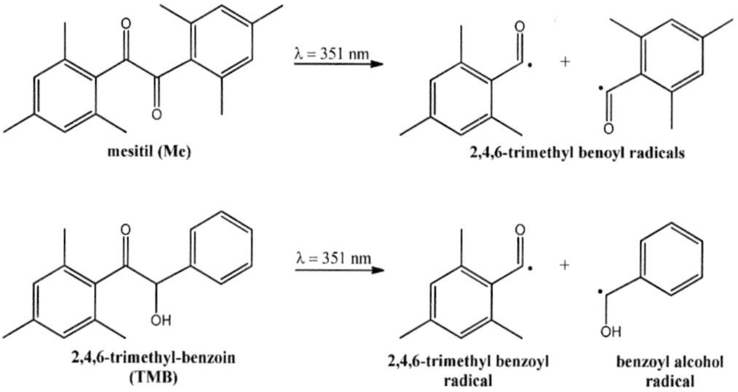

Figure 5.1: Photolytic decomposition of **mesitil (Me)** and **2,4,6-trimethylbenzoin (TMB)** into their radical fragments after irradiation in the UV

As has been demonstrated by a previous study of benzoin-like type-I initiators, the combination of PLP-ESI-MS and TA experiments serves as a powerful tool for a quantitative analysis of the initiation capability of identical radical fragments originating from disparate original photoinitiator molecules.[195] Figure 5.1 shows the photolytic decomposition of 2,4,6-trimethyl-benzoin (TMB) and mesitil (Me) into their radical fragments after ICT excitation in the UV. A distinct difference of incorporation ability of the 2,4,6-trimethyl benzoyl radical was detected. Even though in the case of Me two 2,4,6-trimethyl benzoyl radicals are formed, less initiation efficiency was found compared to TMB. In other words, identical radical fragments are not automatically featured by the same initiation efficiency if their originating molecules are different. Furthermore, it turned out that less the extinction coefficients and related excitation probability are the determining factor than the quantum yield of the underlying ISC process, and thus the ability of radical formation. For Me it was found that competitive relaxation processes, e.g. IC, diminish relaxation via ISC and finally lead to a low reaction yield of α-cleavage.

In the present work, we went one step further and compared for the first time different radical fragments in order to get deeper insights into the decisive factors for an efficient polymerization reaction. The initiation efficiencies of the radicals of the three type-I photoinitiators 2-methyl-4'-(methylthio)-2-morpholinopropiophenone (MMMP), benzoin (2-hydroxy-1,2-diphenylethanone, Bz) and 4-methyl-benzoin (4MB) are determined by PLP-ESI-MS experiments at an excitation wavelength of 351 nm employing mixtures which contain the monomer methyl methacrylate (for further experimental details see reference 178). The photolytic decomposition via α-cleavage reaction is displayed in figure 5.2. As can be seen, the benzyl-type fragments T, B and MB only differ in their substituents on the benzoyl ring in *para* position. MMMP is widely used in industrial applications and its generated radicals are characterized by nearly equal initiation abilities.[196] A quantitative determination, however, has not been carried out yet. Table 5.1 contains the initiation efficiencies of the investigated radical fragments which were obtained from post-mortem analysis and normalized to the concentration of the photoinitiators.

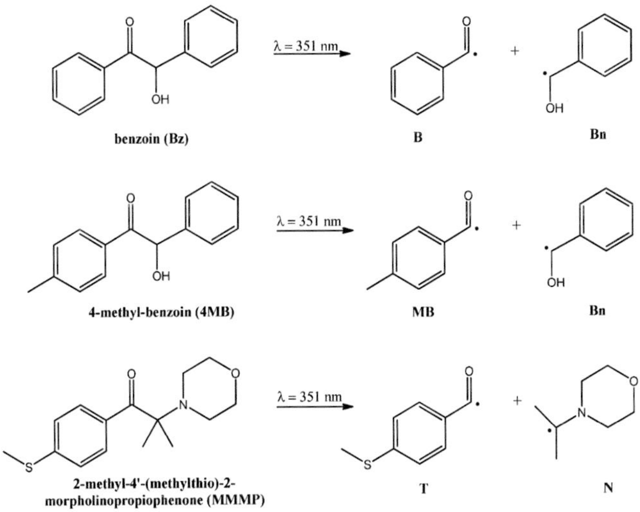

Figure 5.2: Photolytic decomposition of **benzoin (Bz)**, **4-methyl-benzoin (4MB)** and **2-methyl-4'-(methylthio)-2-morpholinopropiophenone (MMMP)** into their radical fragments after irradiation in the UV. The benzyl-type radicals are abbreviated by B, MB and T, whereas the benzoyl alcohol radical and the morpholino radical are named with Bn and N, respectively.

The highest value was found for the benzoyl fragment B of Bz compared to the fragment MB from 4MB and the two fragments N and T from MMMP. Since the Bn fragment is produced by both Bz and 4MB, it could not be compared. In addition, the N and T fragments of MMMP reveal higher initiation efficiencies than the MB radical.

To elucidate the very first step of the polymerization process, and thus the excited state dynamics leading to α-cleavage in the T_1 state, steady state as well as fs time-resolved spectroscopy experiments were performed. Since the photoinitiators were excited at 351 nm in the PLP-ESI-MS experiments, the same wavelength was chosen to pump the molecules in the TA measurements. Moreover, methyl-isobutyrate (MIB) was employed as solvent due to its structural similarity to the monomer used in the polymerization experiments. In order to get deeper insights into the excited state properties of the investigated photoinitiators, quantum chemical calculation at the (TD)-DFT level of theory were performed. The following sections provide a detailed discussion of the obtained experimental and theoretical findings. A concluding comparison regarding the results derived from PLP-ESI-MS experiments is given at the end of this chapter.

Table 5.1: Initiation efficiencies of compared radical fragments derived from PLP-ESI-MS experiments.[178] The values are normalized to the concentrations of the photoinitiators. For more details, see text.

Compared radical fragments	Initiation efficiencies obtained from PLP-ESI-MS experiments
B:MB	1:0.86
B:T	1:0.63
B:N	1:0.45
T:MB	1:0.70
N:MB	1:0.92

5.2   Steady state spectra and calculated vertical transitions

Figure 5.3 shows a direct comparison of experimental extinction coefficients of the three selected photoinitiators Bz, 4MB and MMMP in MIB. As a result of the similar molecular structures of Bz and 4MB, their steady state spectra exhibit comparable features. For both initiators a weak absorption band centered at 320 nm and several stronger ones below 250 nm (not all shown here) can be identified. Considering Bz, two weak shoulders around 284 and 292 nm are found, whereas 4MB reveals a shoulder centered at 257 nm. Conversely, a broad absorption band at 304 nm is found which shows a small shoulder close to 350 nm. Further transitions are centered below 250 nm and are not depicted in figure 5.3. Since the initiators were excited at 351 nm in the PLP-ESI-MS experiments, the same wavelength was chosen for TA experiments. The inset graph allows for a closer look at the area around the pump wavelength. Extinction coefficients (in units of $10^3 L/(mol \cdot cm)$) at 351 nm are found to be 0.354 (MMMP), 0.245 (4MB) and 0.110 (Bz), and thus are very low.

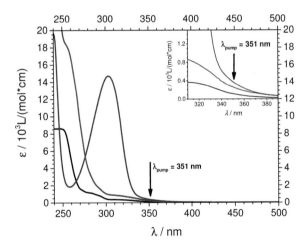

Figure 5.3: Wavelength dependent extinction coefficients of **Bz** (black), **4MB** (red) and **MMMP** (blue) in MIB. The excitation wavelength of time-resolved experiments was at 351 nm for all photoinitiators. The inset graph depicts the area between 310 and 390 nm in more detail.

To obtain deeper insights into the character of the excited states involved in the excitation at 351 nm, calculations at the (TD)-DFT level of theory were performed. Figure 5.4 shows selected calculated vertical singlet-singlet as well as singlet-triplet transitions at the Franck-Condon region in combination with the experimental extinction coefficients for all initiators (for numerical values see appendix C table C.4–C.6). In doing so, all calculations are related to the corresponding optimized electronic ground states S_0 whose geometries are inset in the graphs (for Cartesian Coordinates, see appendix C table C.1–C.3). Taking into account previous studies of benzoin-like initiators,[195, 197] structures characterized by *trans* position with regard to the hydroxyl and carbonyl group have been identified as the most stable and relevant forms for α-cleavage from triplet states in polar solvents. Thus, conceivable optimized structures with *cis* arrangement for Bz and 4MB are not included in this study. Comparing theoretical and experimental findings, calculated vertical singlet-singlet transitions are in reasonable agreement with the UV-Vis data revealing a slight red shift. Consequently, it can be concluded that excitation at 351 nm corresponds to transition into the first excited singlet state (S_1) for all molecules. For MMMP, however, additional contributions of higher excited singlet states (at least S_2) are assumed, considering the proximity of the strong absorption band (304 nm) and the shoulder around 350 nm. Furthermore, high oscillator strengths around 304 nm as well as the FWHM of the pump pulses about 6-10 nm support this assumption.

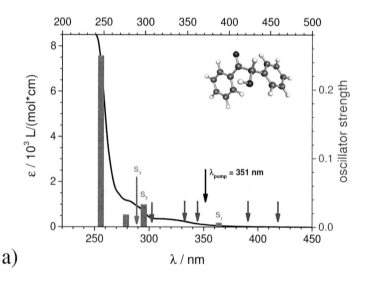

a)

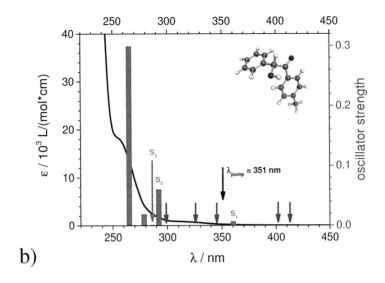

b)

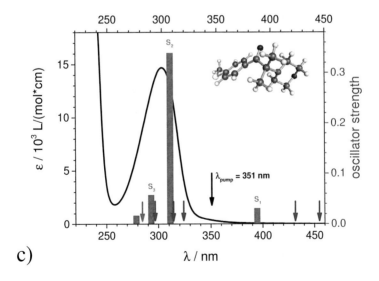

c)

Figure 5.4: Measured absorption spectra of **Bz (a)**, **4MB (b)** and **MMMP (c)** in MIB (black lines). Calculated oscillator strengths of selected singlet-singlet transitions are displayed as grey bars (right axis). Singlet-triplet transitions are marked by blue arrows. The excitation wavelength (λ_{pump}) of time-resolved experiments was at 351 nm. Optimized geometries of the ground state are inset in the graphs.

As inspection of molecular orbitals indicates, excitation into the first excited singlet state is associated with transition from the HOMO into the LUMO for all three photoinitiators. The involved orbitals are shown in figure 5.5. For Bz, the HOMO is specified by electron distribution over the phenyl ring as well as the lone pair of the carbonyl oxygen. In contrast, the main contribution of the LUMO is described by π^*-MO implying the benzoyl ring and the carbonyl oxygen being in line with the literature.[195, 197] Since Bz and 4MB have similar molecular structures, the results obtained for Bz can be sufficiently extrapolated to 4MB (figure 5.5 b)). For MMMP (figure 5.5 c)), contributions to the HOMO are given by a lone pair character of the carbonyl oxygen and, as a consequence of the absence of a phenyl ring, by a lone pair character of the nitrogen atom of the morpholino group. Moreover, the LUMO of MMMP is characterized by π^*-MO implying the benzoyl ring and the carbonyl oxygen as found for Bz and 4MB agreeing well with a previous study.[198] In summary, excitation at 351 nm is described by ICT transition from the hydroxybenzyl and morpholino moiety, respectively, to the benzoyl fragment into the first excited singlet state S_1. The latter reveals $^1(n\pi)^*$ character for all three photoinitiators.

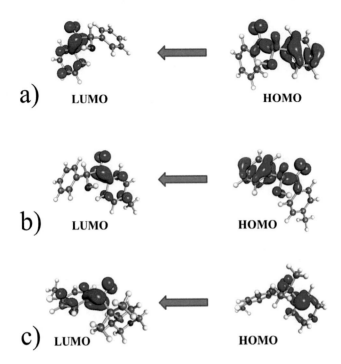

Figure 5.5: HOMO and LUMO orbitals for **Bz (a)**, **4MB (b)** and **MMMP (c)** involved in the ICT excitation into the first excited singlet state.

5.3 Transient absorption spectroscopy

After excitation into S_1, the evolution of the excited state depopulation was recorded using a white light continuum in the range between 435 and 600 nm. As displayed in figure 5.6 a) and b), the TA spectra of Bz and 4MB exhibit a similar feature. For both molecules, dominant absorption is found in a wavelength range between 430 and 540 nm which decays on a ps time scale and results in constant residual ΔOD values. On the contrary, two distinguishable absorption bands centered at about 520 and 464 nm can be identified for MMMP (figure 5.6 c)). Thereby, the band around 520 nm directly occurs at time zero and further decays within the accessible time window of 1800 ps, whereas the band at 464 nm appears considerably later. Apparently, the decrease of the first band as well as the increase of the second one takes place on a similar time scale. Thus, it can be concluded that depopulation of the excited species observed around 520 nm directly leads to population of another excited species which reveals high absorption cross sections at 464 nm.

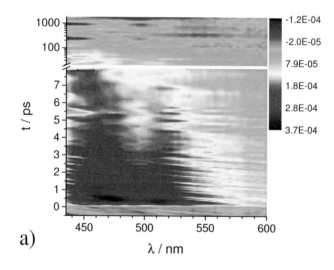

a)

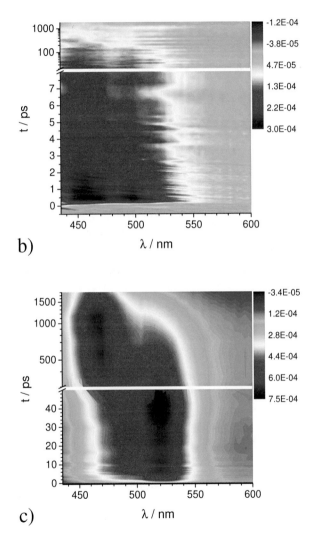

Figure 5.6: TA profiles of **Bz (a)**, **4MB (b)** and **MMMP (c)** in MIB at a pump wavelength of 351 nm and probed using a white light continuum between 435 and 600 nm.

To obtain a more quantitative picture, the TA spectra were analyzed by multiexponential fit functions as described in section 3.4. In doing so, best results were achieved by a sum of three exponentials for Bz and 4MB using the global fit routine. As can be seen from table 5.2, biexponential decay is found for Bz and 4MB by fitting the spectral region between 441 and 552 nm (Bz) as well as between 441 and 540 nm (4MB) resulting in the time constants τ_1 and τ_2. Comparing the temporal evolution for both initiators, the values for τ_1 are well comparable, while the second time constant is substantially lower for 4MB than for Bz (difference of approximately 30-100 ps). Furthermore, the observed residual absorption is associated with τ_3 and lies outside of the experimental time window. Since τ_3 is undefined, the determination of τ_2 is more inaccurate than the determination of τ_1. As a result of several tests, the error margin of τ_2 is found to be higher than the standard deviations from the fitting procedure. Therefore the value of τ_2 is not specified and a range of time constants is given instead. For MMMP, global analysis of the spectral region between 435 and 549 nm yielded adequate description of TA traces with a fourfold exponential function. Biexponential increase of ΔOD is found within the first 100 ps which is associated with τ_1 and τ_2. Moreover, τ_3 describes the decay of the first absorption band (around 520 nm) as well as the rise of the second one (around 464 nm). The fourth time constant τ_4 is required to model the decay for long delay times and lies outside of the experimental time window. All obtained time constants are listed in table 5.2. As for Bz and 4MB, the error margin of τ_3 is found to be larger than the standard deviation of the fitting procedure, and thus a range of time constants is given for τ_3. Corresponding DADS spectra of all photoinitiators are shown and discussed later (see section 5.4).

However, as a consequence of the small extinction coefficients at 351 nm as well as of the employed pump wavelength generation, the intensity of the recorded TA spectra was very low. In order to obtain higher ΔOD values, further experiments were performed at an excitation wavelength of 325 nm. These data are characterized by a substantially increased quality of the TA experiments resulting in a better signal-to-noise ratio which allows for an improved analysis with respect to the time constants. All TA spectra recorded at 325 nm and obtained time constants are listed in appendix C figure C.1 and table C.7. Since the observed TA features are well comparable for Bz and 4MB at 351 and 325 nm, excitation into the same excited state is very likely. Additionally, this assumption is supported by the steady state spectra as well as the computed singlet transitions (see figure 5.4). In comparison to the data recorded at 351 nm, slightly lower time constants are found due to transition into another

Franck-Condon region of S_1. On the contrary, excitation at 325 nm leads to an increase of contributions of higher excited singlet states to TA spectra for MMMP (considering figure 5.4 c)). As a result, a spectral shift of first absorption band from 520 to 500 nm was observed finally resulting in lower time constants at 325 nm. In summary, the obtained results at 325 nm support the findings of the TA data taken at 351 nm except of slight differences due to higher excess energy.

Table 5.2: Time constants of **Bz**, **4MB** and **MMMP** obtained by global fit routine in MIB at 351 nm. The given errors are standard deviations derived from the fitting procedure. For τ_2 (Bz and 4MB) and τ_3 (MMMP) a range of time constants is given. For more details, see text.

photoinitiator	τ_1/ps	τ_2/ps	τ_3/ps	τ_4/ps
Bz	3.2 ± 0.2	(50-80)	> 1800	---
4MB	2.5 ± 0.1	(80-180)	> 1800	---
MMMP	1.9 ± 0.1	70 ± 7	(650-850)	> 1800

5.4 Relaxation pathways of the photoinitiators after intramolecular charge transfer excitation

The following section aims at a detailed discussion and comparison of the results obtained for all three photoinitiators. As a consequence of their similar molecular structure as well as their related TA features, Bz and 4MB are discussed together, whereas the relaxation pathways of MMMP are examined afterwards. A concluding comparison of post-mortem analysis and TA data is provided in section 5.5.

5.4.1 Benzoin (Bz) and 4-methyl-benzoin (4MB)

Taking into account previous studies of Bz[187, 197] and similar type-I photoinitiator systems[199-201], time constants for α-cleavage reaction were found on a ps time scale. Thus, after ICT excitation into S_1 the population of T_1 via ISC and its depopulation via radical formation can be monitored within the experimental time window of 1800 ps. In analogy to our earlier study of Bz in methanol[195], τ_3 (corresponding to the observed remaining absorption for long delay times) is assigned to the absorption of reaction products generated by α-cleavage. The preceding depopulation of T_1 can be associated with the second time constant τ_2. In comparison to reference 195, the obtained value of 50-80 ps for Bz in MIB is considerably larger than 14.3 ps found in methanol. This finding points to a strong solvent dependence of the α-cleavage reaction in line with a previous study on solvent effects on the photochemistry of Bz conducted by Lipson *et al.*[197] As can be seen from table 5.2, larger time constants for α-cleavage are found for 4MB compared to Bz in MIB. In addition, depopulation of S_1 via ISC is connected with τ_1 being in the order of few ps, and thus in good agreement with the value derived for Bz in methanol $((1.2 \pm 0.2) \text{ ps})$[195] and with values observed for similar type-I photoinitiator molecules.[202] Considering the $^1(n\pi^*)$ character of S_1, such states can easily undergo ISC processes into $^3(\pi\pi^*)$ states according to the El-Sayed selection rules.[156, 157] Since the smallest energy gap was found between S_1 and T_2 (0.238 eV for Bz; 0.268 eV for 4MB) revealing at least partly $^3(\pi\pi^*)$ character, ISC into T_2 is very likely for Bz and 4MB. Subsequently, rapid vibrational relaxation leads to population of the T_1 state which is sufficiently long-lived that α-cleavage can occur. As has already been mentioned in the previous chapter (section 4.2.4.2), IC into the electronic ground state is a competitive relaxation process and cannot be excluded for τ_1. Hence, τ_1 has to be reckoned as a lower limit for the underlying ISC process. The proposed relaxation pathways of Bz and 4MB are summarized in a Jabłoński-type sketch shown in figure 5.7.

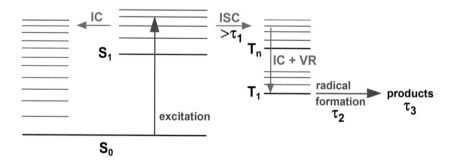

Figure 5.7: Jabłoński-type sketch of the proposed relaxation mechanism of **Bz** and **4MB**. After ICT excitation into S_1 ISC into the upper triplet manifold is found to be $>\tau_1$. Therein, rapid vibrational relaxation leads to population of the T_1 state which is sufficiently long-lived that radical formation via α-cleavage (τ_2) takes place. The absorption of generated radical products can be related to τ_3.

In order to investigate the influence of the different relaxation steps on the TA data, DADS spectra are shown in figure 5.8. The exponential decay of S_1 corresponds to amplitude A, whereas depopulation of T_1 and the absorption of radical products are associated with the amplitudes B and C, respectively. For the amplitude A, a similar trend is observed for Bz and 4MB. Moreover, the evolution of the amplitudes B and C are comparable for both photoinitiators, yet exhibiting a higher interspace for Bz. To get a more detailed insight relative amplitudes (see section 3.4) were calculated for three selected probe wavelengths (477, 489 and 501 nm) in the vicinity of the peak absorption and are summarized in table 5.3. Integration over the entire absorption band instead of three probe wavelengths would provide more reliable values, but caused by limitations of the used WL generation (see section 3.3 and 3.5) the relevant spectral signature of the photoinitiators could only be partially recorded (no TA data below 435 nm accessible). Additionally, an increasing number of excited state contributions (considering Figure 5.4 a) and b)) to the observed ESA complicate an unambiguous assignment to specific singlet or triplet transitions. Furthermore, required information, e.g. absorption cross sections, of the electronically excited species is not easily accessible.

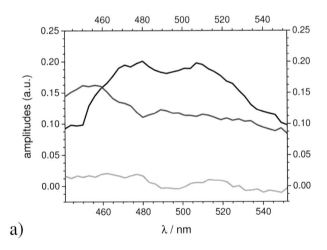

a)

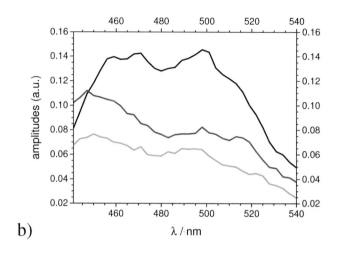

b)

Figure 5.8: DADS spectra for **Bz (a)** and **4MB (b)** in **MIB** at a pump wavelength of 351 nm. Amplitude A is displayed by black, amplitude B by red and amplitude C by green lines. For more details, see text.

As one can see from table 5.3, the first amplitude A_{rel} shows the highest contribution to the TA for both Bz and 4MB. Comparing the second and the third amplitudes, C_{rel} plays only a minor role in the case of Bz. For 4MB, B_{rel} and C_{rel} are nearly equal pointing to a higher contribution of radical products of α-cleavage than for Bz. On the assumption of similar absorption properties of the initially excited singlet states and the radical products, estimation of the ability of radical formation is provided by the normalized TA profiles at the selected probe wavelengths (477, 489 and 501 nm) as displayed in figure 5.9. In doing so, the resulting ΔOD (a.u.) values of 0.08 for Bz and 0.27 for 4MB indicates that 4MB reveals a ~3.4 times higher capability towards radical formation than Bz. In general, the efficiency of radical formation is directly linked to the efficiency of the ISC process as well as further relaxation processes (as e.g. IC) which lead to depopulation of the T_1 state. A quantitative determination of the ISC efficiency, however, was not possible because TA data do not allow for the required quantitative examination of competitive reactions. Assuming that additional IC processes are negligible, a quantum yield of $\Phi_{radical} = 1$ is given for radical formation originating from T_1. Then, the estimated value of ~3.4 can be directly related to the relationship of the ISC efficiencies between Bz and 4MB. In summary, it can be concluded that a higher ability of radical generation is found for 4MB in comparison to Bz. A similar behavior is also observable for TA data recorded at 325 nm resulting in a ~2.5 times higher capability of radical formation for 4MB. The detailed analysis of TA data at 325 nm (DADS spectra, relative amplitudes and normalized TA traces) is shown in appendix C figure C.2, C.3 and table C.8.

Table 5.3: Relative amplitudes (according to section 3.4) of **Bz** and **4MB** at a pump wavelength of 351 nm and three selected probe wavelengths as labeled.

photoinitiator		A_{rel}	B_{rel}	C_{rel}
Bz	477 nm	0.59	0.35	0.06
	489 nm	0.59	0.40	0.01
	501 nm	0.62	0.37	0.01
4MB	477 nm	0.49	0.29	0.22
	489 nm	0.49	0.28	0.23
	501 nm	0.51	0.28	0.21

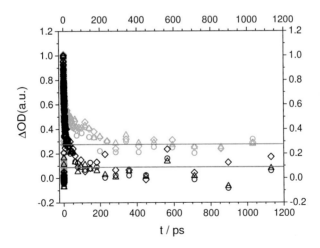

Figure 5.9: Normalized TA traces of **Bz (black)** and **4MB (green)** in MIB at the three selected probe wavelengths 477 nm (squares), 489 nm (triangles) and 501 nm (circles) and at a pump wavelength of 351 nm. ΔOD (a.u.) values of 0.08 for Bz and 0.27 for 4MB are indicated by grey lines. For more details, see text above.

5.4.2   2-methyl-4'-(methylthio)-2-morpholinopropiophenone (MMMP)

Considering section 5.2, excitation at 351 nm corresponds to an ICT transition into S_1 and additionally (at least) into S_2. Furthermore, the occurrence of the two spectral and temporal separable absorption bands (figure 5.6 c)), as well as the determination of a fourth time constant, implies a different relaxation behavior compared to Bz and 4MB. As already demonstrated by a ps time-resolved study of Morlet-Savary et al.,[198] the absorption band around 520 nm is associated with singlet-singlet absorption of S_1. In contrast, the band centered at 464 nm is caused by absorption originating from the first excited singlet state T_1. As one can see from figure 5.10, the decay of the singlet absorption band as well as the rise of the triplet absorption band are directly linked to the positive and negative signature of amplitude C, and thus related to the corresponding time constant τ_3. Therefore, the obtained value of (650-850) ps is assigned to depopulation of S_1 via ISC and agrees well with the value of ~700 ps derived by Morlet-Savary et al.[198] An evaluation of the TD-DFT calculations shows that the smallest energy gap (0.268 eV) was found between S_1 and T_2 which reveals at least partly $^3(\pi\pi^*)$ character. Hence, ISC of MMMP most likely takes place from S_1 into T_2 in analogy to Bz and 4MB, but with a considerably higher time constant.

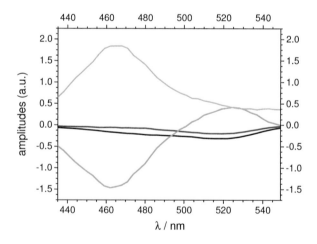

Figure 5.10: DADS spectrum for **MMMP** in **MIB** at a pump wavelength of 351 nm. Amplitude A is displayed by black, amplitude B by red, amplitude C by green and amplitude D by cyan lines. For more details, see text above.

As a consequence, the relaxation process corresponding to τ_4 is associated with the lifetime of the first excited triplet state T_1 and α-cleavage reaction resulting in a positive signature of the related amplitude D. Since τ_4 lies outside of the accessible time window, radical products cannot be observed on TA spectra in contrast to Bz and 4MB. Moreover, the increase of TA within the first 100 ps assigned to τ_1 and τ_2 is displayed by a slight negative signature of the amplitudes A and B. Observation of the biexponential rise indicates the contribution of several different steps, e.g. vibrational state relaxation within S_1 and S_2 (due to additional excitation into S_2), IC from S_2 into S_1 or S_0 as well as ISC from S_2 to the triplet manifold. Thereby, vibrational relaxation is provided by the pump wavelength dependence of τ_1 and τ_2 considering recorded TA spectra and obtained time constants at 325 nm (appendix C). In addition, ISC and IC processes caused by excitation into higher singlet states are assumed in the study of Morlet-Savary et al.,[198] but not further specified. Although TA experiments were performed with a time resolution of ~100 fs, it is not possible to disentangle the different underlying processes of τ_1 and τ_2. The proposed relaxation pathway of MMMP is sketched in a Jabłoński-type diagram shown in figure 5.11.

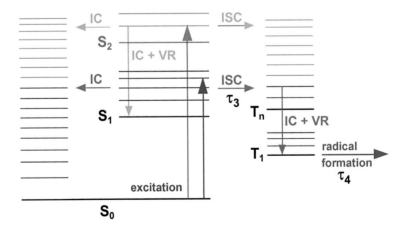

Figure 5.11: Jabłoński-type sketch of the proposed relaxation mechanism of **MMMP**. Excitation at 351 nm leads to transition into S_1 and partly S_2. ISC into the upper triplet manifold is assigned to the time constant τ_3. Rapid vibrational relaxation therein leads to population of T_1 which is sufficiently long-lived that radical formation via α-cleavage (τ_4) occurs. Radical products are not observed within the accessible time window of 1800 ps. Displayed relaxation processes caused by excitation into S_2 are associated with τ_1 and τ_4, but are not further specified.

As mentioned above, α-cleavage products cannot be observed on TA spectra and, as a result, estimation of the ability of radical formation by normalized TA traces is not possible in the same way as for Bz and 4MB. Furthermore, due to the spectral differences of the singlet and triplet absorption bands, relative amplitudes cannot be determined at one probe wavelength for MMMP. However, the number of generated molecules in S_1 and T_1 can be estimated from the maximum absorption of the two bands, assuming that contributions of higher triplet states are negligible and absorption properties are comparable for both excited species. In doing so, ΔOD (a.u.) values of 0.72 (singlet band, figure 5.12 a)) and 0.68 (triplet band, figure 5.12 b)) are obtained. The relationship between these values (~0.94) can be reckoned as a measure of the ISC efficiency originating from S_1 or, in other words, population of T_1 is calculated to be about 94%, and thus very high. Assuming a quantum yield of $\Phi_{radical} = 1$ as for Bz and 4MB, the calculated relation also postulates a high capability of radical formation for MMMP. An analog evaluation of the TA data recorded at 323 nm results in a T_1 population of about 95% (the detailed estimation of this value is depicted in appendix C figure C.4 and C.5). In summary, it can be concluded that a higher ability of radical generation is found for 4MB and most likely also for MMMP in comparison to Bz.

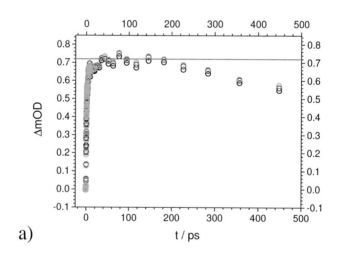

a)

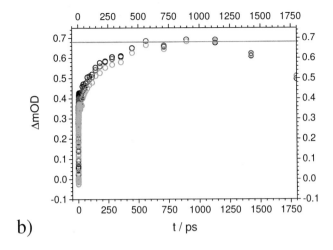

b)

t / ps

Figure 5.12: Singlet absorption band of **MMMP (a)** at 525 nm (black), 522 nm (red) and 519 nm (green) as well as **triplet absorption band (b)** at 471 nm (black), 468 nm (red) and 465 nm (green). The pump wavelength was at 351 nm. ΔOD (a.u.) values of 0.72 (a) and 0.68 (b) are indicated by grey lines. For more details, see text above.

5.5 Conclusion

The following section aims at comparing results from steady state and fs time-resolved spectroscopy as well as PLP-ESI-MS experiments. Concerning only the extinction coefficients ε at 351 nm, one would expect a higher initiation ability in the order of MMMP ($\varepsilon = 0.354 \cdot 10^3 L/(mol \cdot cm)$), 4MB ($0.245 \cdot 10^3 L/(mol \cdot cm)$) and Bz ($0.110 \cdot 10^3 L/(mol \cdot cm)$) due to a higher excitation probability. However, it is obvious that the magnitude of ε does not correlate with the initiation efficiencies determined from the post-mortem analysis (table 5.4 left hand column). Thus, it can be stated that an inspection of extinction coefficients is not sufficient for predicting initiation abilities of the investigated radical fragments as also observed in a previous study of similar initiator molecules.[195] Since the efficiency ratios derived from PLP-ESI-MS experiments are normalized to the concentration of the photoinitiators, these relationships have to be made independent of the excitation probabilities. In doing so, the efficiency values are multiplied by the corresponding ratios of ε and are summarized in table 5.4. The initiation efficiencies scaled to both the concentration and the respective excitation probabilities show the highest value for the benzoyl fragment B of Bz compared to the fragment MB from 4MB and the two fragments N and T from MMMP. As a result of the lower extinction coefficients of 4MB compared to MMMP, the ratios of the initiation efficiencies (T:MB and T:N) change by scaling them to the same number of excited molecules. However, comparing the estimated ability of radical generation from the TA data (see section 5.4.1 and 5.4.2), a higher tendency for α-cleavage is found for 4MB with respect to Bz. Furthermore, despite the more complex relaxation behavior due to partial excitation into (at least) the second excited singlet state, a higher ability of radical formation is found for MMMP compared to Bz, too. Hence, unlike to the previous study on identical radical fragments originating from different initiator molecules (comparison of a 2,4,6-trimethylbenzoyl fragment generated from mesitil and 2,4,6-trimethylbenzoin as mentioned in the introduction part 5.1) the capability of radical formation seems not to be the crucial step for efficient photoinduced initiation if radicals with different molecular structures are investigated. A further aspect which may influence the efficiency of α-cleavage reaction is the nature of the T_1 state, as discussed in several earlier studies.[197, 199, 203] Thereby, T_1 states with $^3(n\pi^*)$ character are supposed to lead to higher radical generation than those revealing $^3(\pi\pi^*)$ character. Orbital analyses of Bz, 4MB and MMMP, however, reveal only slight differences between the T_1 states. All photoinitiators are characterized by at least partly $^3(n\pi^*)$ nature indicating that the observed discrepancies of radical formation are apparently caused by other factors. Additionally, different conformations due to internal rotation during ISC can affect the α-cleavage reaction,[1, 204, 205] but information about rotation around the C(O)–C(OH) (for Bz and 4MB) and C(O)–C(Me)$_2$ (for MMMP) bond are not provided by the TA data. Nevertheless, an analog behavior can be assumed for Bz and 4MB due to their similar molecular structures.

As demonstrated in this study, neither inspection of extinction coefficients nor determination of the ISC efficiency is sufficient to explain the obtained initiation efficiencies from the post-mortem analysis by comparing radicals with different molecular structures from disparate source molecules. Therefore, the varieties of photophysical properties seem to be less important than the reactivity of the generated radical fragments itself. Comparing the B, MB and T fragments, the radicals only differ in their substituents on the benzoyl ring in *para* position. The initiation efficiencies normalized to ε (see table 5.4 right hand column) point to a decrease of the initiation ability for both substituted fragments MB and T most likely due to an increase of steric effects. The influence of sterical hindrance regarding incorporation probabilities of radicals has already been demonstrated by Voll *et al.* by benzoyl fragments with a various number of methyl substituents.[91] As a consequence of its different molecular structure, the N fragment of MMMP cannot be easily compared, but the occurrence of a six-membered ring and the two branched methyl groups suggest that steric effects play a decisive role, too. Thus, the divergent reactivity of the generated radicals regarding the reaction onto the vinylic double bond of the employed monomers in the PLP-ESI-MS experiments cannot be neglected. As previously mentioned, information about photophysical properties and post-mortem analysis allow for quantification of initiation efficiencies when comparing the same fragments from different source molecules.[195] Additional information, however, e.g. the addition rate coefficients of the radicals for addition reactions with respect to the C-C double bond are required for different fragments. In summary it can be concluded, that a benzoin-like structure serves as an appropriate framework for useful photoinitiators. Thereby, substitution should enhance the interplay between minimal sterical hindrance and maximal ISC efficiency in order to increase the initiation efficiency. Moreover, different electron donating or withdrawing substituents change the strength of the ICT character of the initially excited state. In this way, also the efficiency of subsequent relaxation processes, as e.g. IC, ISC and radical formation can be influenced. Thus, a further systematic investigation of substituent effects on the excited state properties and in particular the time range for efficient ISC, the reactivity and the stabilization of radical fragments, and finally the incorporation probabilities is part of future necessary work.

Table 5.4: Initiation efficiencies of radical fragments normalized to the concentration of the photoinitiators are listed in the left hand column. By multiplying the values unequal one by the corresponding ratio of extinction coefficients, initiation efficiencies scaled to the same number of excited molecules are obtained.

Compared radical fragments	Initiation efficiencies obtained from PLP-ESI-MS experiments	Ratio of extinction coefficients at 351 nm	Initiation efficiencies scaled to extinction coefficients
B:MB	1:0.86	$\dfrac{\varepsilon(Bz)}{\varepsilon(4MB)} = 0.449$	1:0.39
B:T **B:N**	1:0.63 1:0.45	$\dfrac{\varepsilon(Bz)}{\varepsilon(MMMP)} = 0.451$	1:0.28 1:0.20
T:MB **N:MB**	1:0.70 1:0.92	$\dfrac{\varepsilon(MMMP)}{\varepsilon(4MB)} = 1.445$	1:1.01 1:1.33

A. Appendix for Chapter 3

Concentration dependent steady state spectroscopy

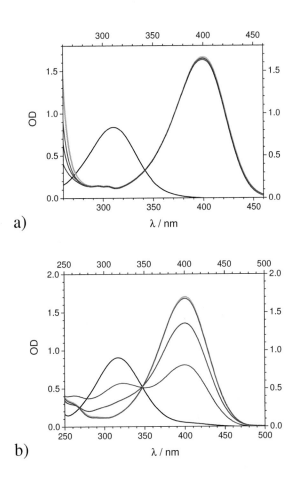

a)

b)

Figure A: Steady state spectroscopy of *p*-**NP** in chloroform (a) and water (b) with different concentrations of DBU and NaOH, respectively. The ratios (nitrophenol:base) 1:0.5, 1:1, 1:1.5, 1:2 and 1:2.5 are displayed as grey, blue, red, green and cyan lines. OD without base is shown as black lines. Complete deprotonation is assumed at the ratio (nitrophenol:base) of 1:1.5 and 1:1 for NaOH and DBU. Analog concentration dependence was found for *m*-NP and *o*-NP (not shown).

B. Appendix for Chapter 4

B.1 Geometry optimization

B.1.1 Ground states

Table B.1: Cartesian coordinates in Å of the ground state minimum of *o*-**NP** optimized by DFT.

Atom	x	y	z
C	-1.03734	1.01485	0.00000
C	-1.19536	-1.68658	0.00000
C	1.10833	-3.05570	0.00000
C	3.42444	-1.79784	0.00000
C	3.54761	0.87566	0.00000
C	1.31627	2.27210	0.00000
O	-3.37641	-2.94935	0.00000
H	0.98956	-5.13300	0.00000
H	5.18534	-2.91211	0.00000
H	5.38806	1.84703	0.00000
H	1.31169	4.35001	0.00000
N	-3.32774	2.53272	0.00000
O	-5.41881	1.38069	0.00000
O	-3.17647	4.84489	0.00000
H	-4.73917	-1.58337	0.00000

Table B.2: Cartesian coordinates in Å of the ground state minimum of *m*-**NP** optimized by DFT.

Atom	x	y	z
C	-2.76250	0.64721	0.00000
C	-2.81408	-2.02427	0.00000
C	-0.54979	-3.39930	0.00000
C	1.80180	-2.16385	0.00000
C	1.80173	0.48290	0.00000
C	-0.42593	1.91809	0.00000
O	-5.02005	1.88058	0.00000
H	-4.66197	-2.98138	0.00000
H	-0.61830	-5.48105	0.00000
H	3.60495	-3.19206	0.00000
H	-0.28683	3.99433	0.00000
N	4.25899	1.85271	0.00000
O	4.18996	4.18297	0.00000
O	6.20244	0.57019	0.00000
H	-4.72041	3.71290	0.00000

Table B.3: Cartesian coordinates in Å of the ground state minimum of p-**NP** optimized by DFT.

Atom	x	y	z
C	-2.73409	1.55464	0.00000
C	-2.80182	-1.11782	0.00000
C	-0.55133	-2.49438	0.00000
C	1.77046	-1.20064	0.00000
C	1.87270	1.44984	0.00000
C	-0.38331	2.82710	0.00000
O	-4.98196	2.79181	0.00000
H	-4.65092	-2.07163	0.00000
H	-0.54725	-4.57214	0.00000
H	3.72460	2.39185	0.00000
H	-0.32827	4.91306	0.00000
N	4.15038	-2.65706	0.00000
O	6.15109	-1.45431	0.00000
O	3.99102	-4.98504	0.00000
H	-4.68127	4.62472	0.00000

Table B.4: Cartesian coordinates in Å of the ground state minimum of o-**NP**$^{\ominus}$ optimized by DFT.

Atom	x	y	z
C	-2.52616	0.40240	0.00000
C	-2.27696	-2.37385	0.00000
C	0.00311	-3.65396	0.00000
C	2.33909	-2.31101	0.00000
C	2.27286	0.32002	0.00000
C	-0.03006	1.68529	0.00000
H	-4.09586	-3.39982	0.00000
H	0.01805	-5.74755	0.00000
H	4.16040	-3.32804	0.00000
H	4.01835	1.45211	0.00000
N	0.18277	4.41368	0.00000
O	-4.65744	1.42457	0.00000
O	-1.76296	5.72844	0.00000
O	2.35480	5.38771	0.00000

Table B.5: Cartesian coordinates in Å of the ground state minimum of *m*-NP$^\ominus$ optimized by DFT.

Atom	x	y	z
C	-3.28247	0.95116	0.00000
C	-3.12242	-1.81514	0.00000
C	-0.83518	-3.14221	0.00000
C	1.51972	-1.89794	0.00000
C	1.44393	0.77422	0.00000
C	-0.80007	2.17221	0.00000
H	-4.94406	-2.83869	0.00000
H	-0.86665	-5.23498	0.00000
H	3.33633	-2.89817	0.00000
H	-0.72498	4.24911	0.00000
N	3.88154	2.14510	0.00000
O	-5.33867	2.15396	0.00000
O	3.86235	4.48869	0.00000
O	5.87066	0.89267	0.00000

Table B.6: Cartesian coordinates in Å of the ground state minimum of *p*-NP$^\ominus$ optimized by DFT.

Atom	x	y	z
C	-3.23678	1.91650	0.00000
C	-3.11696	-0.85737	0.00000
C	-0.86374	-2.18639	0.00000
C	1.48821	-0.88115	0.00000
C	1.50184	1.80835	0.00000
C	-0.74681	3.14493	0.00000
H	-4.94895	-1.85961	0.00000
H	-0.81368	-4.26755	0.00000
H	3.35069	2.76535	0.00000
H	-0.74902	5.23307	0.00000
N	3.81300	-2.25808	0.00000
O	-5.28201	3.12798	0.00000
O	5.85964	-1.05599	0.00000
O	3.74459	-4.63002	0.00000

B.1.2 Excited singlet states

Table B.7: Optimization of the first excited singlet state of *o*-**NP** failed (no stable minimum structure could be found). Cartesian coordinates in Å of the last calculation step optimized by TD-DFT are shown below.

Atom	x	y	z
C	0.26976	-0.40951	-0.13879
C	0.01863	-0.52840	2.55281
C	2.31654	-0.39284	3.98692
C	4.66060	0.00411	2.75343
C	4.75956	0.37672	0.14917
C	2.45582	0.29996	-1.33204
O	-2.25016	-0.36219	3.44216
H	2.24622	-0.63852	6.05391
H	6.39220	0.00304	3.91069
H	6.55635	0.60326	-0.87658
H	2.56510	0.65683	-3.38204
N	-2.27765	-0.37940	-1.40064
O	-3.72718	1.56988	-0.60417
O	-2.94850	-1.88829	-3.01169
H	-3.46698	1.08535	1.44777

Table B.8: Cartesian coordinates in Å of first excited singlet state minimum of *m*-**NP** optimized by TD-DFT.

Atom	x	y	z
C	0.04998	0.00000	-0.04696
C	-0.09365	0.00014	2.62455
C	2.18827	0.00003	3.92625
C	4.55599	-0.00017	2.62754
C	4.66727	-0.00010	-0.01991
C	2.42441	-0.00006	-1.37786
O	-2.12627	-0.00012	-1.37170
H	-1.94523	0.00039	3.56691
H	2.18515	0.00009	6.00907
H	6.37191	-0.00055	3.64197
H	2.57155	-0.00007	-3.46056
N	7.08768	0.00014	-1.40023
O	6.93095	0.00028	-3.82318
O	9.06448	0.00021	-0.00449
H	-1.75945	-0.00023	-3.20657

Table B.9: Cartesian coordinates in Å of first excited singlet state minimum of *p*-**NP** optimized by TD-DFT.

Atom	x	y	z
C	0.08364	-0.03913	-0.14518
C	0.00375	0.00044	2.53729
C	2.26285	0.08516	3.91846
C	4.55823	0.22748	2.60753
C	4.68690	0.09669	-0.04010
C	2.43383	0.01236	-1.42821
O	-2.16775	-0.10733	1.37694
H	-1.85046	0.00909	3.48238
H	2.29157	0.09611	5.99716
H	6.55490	0.10439	-0.95261
H	2.48451	0.00868	-3.51461
N	7.00872	0.36584	4.10415
O	8.94523	-0.33957	2.79778
O	6.75304	-0.38247	6.41231
H	-1.87610	-0.13777	-3.21488

Table B.10: Optimization of the second excited singlet state of *o*-**NP**$^\ominus$ failed (no stable minimum structure could be found). Cartesian coordinates in Å of the last calculation step optimized by TD-DFT are shown below.

Atom	x	y	z
C	-0.12736	0.33227	-0.02660
C	-0.01123	0.33687	2.66783
C	2.28314	-0.02504	3.95988
C	4.52713	-0.34295	2.60735
C	4.42482	-0.30061	-0.06171
C	2.18422	0.01041	-1.40847
H	-1.79790	0.58744	3.70909
H	2.28760	-0.04659	6.04731
H	6.33928	-0.62300	3.60012
H	6.16783	-0.52634	-1.17603
N	2.30471	0.03026	-4.17576
O	-2.35287	0.61110	-1.08786
O	2.50261	-2.18931	-5.20100
O	3.11590	2.14549	-5.12479

Table B.11: Cartesian coordinates in Å of first excited singlet state minimum of *m*-NP$^\ominus$ optimized by TD-DFT.

Atom	x	y	z
C	-0.05492	0.00020	-0.03662
C	-0.07853	-0.00121	2.65661
C	2.21467	-0.00098	3.95025
C	4.54037	0.00154	2.62191
C	4.59827	-0.00133	-0.01141
C	2.32697	-0.00034	-1.33222
H	-1.92346	-0.00430	3.62454
H	2.24333	-0.00228	6.03907
H	6.37172	0.01106	3.60795
H	2.35920	-0.00427	-3.41032
N	7.06331	-0.00523	-1.41344
O	-2.16269	0.00695	-1.35199
O	6.97860	0.00546	-3.84583
O	9.06205	-0.00525	-0.01911

Table B.12: Optimization of the first excited singlet state of *p*-NP$^\ominus$ failed (no stable minimum structure could be found). Cartesian coordinates in Å of the last calculation step optimized by TD-DFT are shown below.

Atom	x	y	z
C	-0.12736	0.33227	-0.02660
C	-0.01123	0.33687	2.66783
C	2.28314	-0.02504	3.95988
C	4.52713	-0.34295	2.60735
C	4.42482	-0.30061	-0.06171
C	2.18422	0.01041	-1.40847
H	-1.79790	0.58744	3.70909
H	2.28760	-0.04659	6.04731
H	6.33928	-0.62300	3.60012
H	6.16783	-0.52634	-1.17603
N	2.30471	0.03026	-4.17576
O	-2.35287	0.61110	-1.08786
O	2.50261	-2.18931	-5.20100
O	3.11590	2.14549	-5.12479

B.1.3 Excited triplet states

Table B.13: Cartesian coordinates in Å of first excited triplet state minimum of *o*-**NP** optimized by TD-DFT.

Atom	x	y	z
C	-1.03774	-0.00074	-0.96878
C	-1.15329	-0.00209	1.78165
C	1.22658	-0.00118	3.11098
C	3.56729	0.00025	1.83248
C	3.59703	0.00106	-0.79988
C	1.26361	0.00064	-2.20852
O	-3.25980	-0.00336	2.98491
H	1.11239	-0.00186	5.19001
H	5.34610	0.00046	2.91419
H	5.39783	0.00124	-1.84496
H	1.27408	0.00059	-4.28782
N	-3.32704	-0.00514	-2.45072
O	-5.59639	0.01311	-1.18889
O	-3.33230	-0.00917	-4.78585
H	-5.07835	0.00619	0.72120

Table B.14: Cartesian coordinates in Å of first excited triplet state minimum of *m*-**NP** optimized by TD-DFT.

Atom	x	y	z
C	-2.72506	-0.00001	-0.61319
C	-2.86787	0.00007	2.10499
C	-0.59389	-0.00002	3.42647
C	1.75342	-0.00018	2.14103
C	1.85604	-0.00019	-0.57770
C	-0.39410	-0.00028	-1.94917
O	-4.93757	0.00023	-1.85031
H	-4.73038	0.00020	3.02488
H	-0.60716	0.00009	5.50895
H	3.57102	-0.00034	3.14658
H	-0.27219	-0.00065	-4.02917
N	4.21475	0.00019	-1.87378
O	4.17150	0.00034	-4.28090
O	6.19745	0.00040	-0.48252
H	-4.63596	0.00016	-3.69615

Table B.15: Cartesian coordinates in Å of first excited triplet state minimum of *p*-**NP** optimized by TD-DFT.

Atom	x	y	z
C	-2.75074	1.56706	0.00000
C	-2.80892	-1.11403	0.00000
C	-0.58482	-2.50496	0.00000
C	1.79524	-1.21375	0.00000
C	1.86886	1.48863	0.00000
C	-0.39095	2.83488	0.00000
O	-5.01128	2.80562	0.00000
H	-4.65851	-2.06779	0.00000
H	-0.63808	-4.58220	0.00000
H	3.69273	2.48364	0.00000
H	-0.33711	4.92129	0.00000
N	3.98514	-2.55791	0.00000
O	6.28732	-1.69437	0.00000
O	4.25651	-5.00264	0.00000
H	-4.70539	4.63654	0.00000

B.2 Excitation energies

B.2.1 Related to the ground state

Table B.16: Excitation energies of selected singlet and triplet states as well as oscillator strengths of vertical transitions with A' symmetry of *o*-**NP** relative to the optimized structure of 1A' (S_0) at the TD-DFT level of theory.

Singlet states	E_{rel}/eV	Oscillator strength	Triplet states	E_{rel}/eV
2A' (S_1)	3.30	$0.573 \cdot 10^{-1}$	T_1	2.54
1A" (S_2)	3.88	$0.274 \cdot 10^{-5}$	T_2	3.06
2A" (S_3)	4.30	$0.449 \cdot 10^{-3}$	T_3	3.26
3A' (S_4)	4.40	0.187	T_4	3.38
4A' (S_5)	5.42	$0.147 \cdot 10^{-1}$	T_5	3.85

Table B.17: Excitation energies of selected singlet and triplet states as well as oscillator strengths of vertical transitions with A' symmetry of **m-NP** relative to the optimized structure of 1A' (S_0) at the TD-DFT level of theory.

Singlet states	E_{rel}/eV	Oscillator strength	Triplet states	E_{rel}/eV
2A' (S_1)	3.57	$0.363 \cdot 10^{-1}$	T_1	2.72
1A" (S_2)	3.76	$0.870 \cdot 10^{-8}$	T_2	2.95
2A" (S_3)	4.21	$0.296 \cdot 10^{-3}$	T_3	3.21
3A' (S_4)	4.52	0.136	T_4	3.38
4A' (S_5)	5.44	$0.815 \cdot 10^{-1}$	T_5	3.70

Table B.18: Excitation energies of selected singlet and triplet states as well as oscillator strengths of vertical transitions with A' symmetry of **p-NP** relative to the optimized structure of 1A' (S_0) at the TD-DFT level of theory.

Singlet states	E_{rel}/eV	Oscillator strength	Triplet states	E_{rel}/eV
1A" (S_1)	3.79	$0.651 \cdot 10^{-8}$	T_1	2.90
2A' (S_2)	4.16	0.254	T_2	3.00
2A" (S_3)	4.27	$0.334 \cdot 10^{-3}$	T_3	3.25
3A' (S_4)	4.44	$0.537 \cdot 10^{-2}$	T_4	3.76
4A' (S_5)	5.28	$0.492 \cdot 10^{-1}$	T_5	4.00

Table B.19: Excitation energies of selected singlet and triplet states as well as oscillator strengths of vertical transitions with A' symmetry of **o-NP**$^\ominus$ relative to the optimized structure of 1A' (S_0) at the TD-DFT level of theory.

Singlet states	E_{rel}/eV	Oscillator strength	Triplet states	E_{rel}/eV
1A" (S_1)	2.81	$0.175 \cdot 10^{-4}$	T_1	1.91
2A' (S_2)	3.04	0.157	T_2	2.56
2A" (S_3)	3.24	$0.182 \cdot 10^{-2}$	T_3	3.14
3A" (S_4)	3.67	$0.580 \cdot 10^{-2}$	T_4	3.24
4A" (S_5)	3.70	$0.810 \cdot 10^{-6}$	T_5	3.30

Table B.20: Excitation energies of selected singlet and triplet states as well as oscillator strengths of vertical transitions with A' symmetry of *m*-NP$^\ominus$ relative to the optimized structure of 1A' (S$_0$) at the TD-DFT level of theory.

Singlet states	E$_{rel}$/eV	Oscillator strength	Triplet states	E$_{rel}$/eV
2A' (S$_1$)	1.65	$0.145 \cdot 10^{-1}$	T$_1$	1.03
1A" (S$_2$)	2.30	$0.429 \cdot 10^{-5}$	T$_2$	2.22
2A" (S$_3$)	3.05	$0.791 \cdot 10^{-4}$	T$_3$	2.95
3A" (S$_4$)	3.51	$0.174 \cdot 10^{-2}$	T$_4$	3.05
4A" (S$_5$)	3.60	$0.121 \cdot 10^{-2}$	T$_5$	3.07

Table B.21: Excitation energies of selected singlet and triplet states as well as oscillator strengths of vertical transitions with A' symmetry of *p*-NP$^\ominus$ relative to the optimized structure of 1A' (S$_0$) at the TD-DFT level of theory.

Singlet states	E$_{rel}$/eV	Oscillator strength	Triplet states	E$_{rel}$/eV
1A" (S$_1$)	3.21	$0.485 \cdot 10^{-9}$	T$_1$	1.81
2A' (S$_2$)	3.39	0.430	T$_2$	2.96
2A" (S$_3$)	3.74	$0.586 \cdot 10^{-2}$	T$_3$	3.21
3A" (S$_4$)	3.87	$0.102 \cdot 10^{-8}$	T$_4$	3.41
4A" (S$_5$)	3.94	$0.176 \cdot 10^{-6}$	T$_5$	3.43

B.2.2 Related to the first excited triplet state

Table B.22: Excitation energies of selected triplet states and corresponding oscillator strengths (vertical transitions) of **m-NP** relative to the optimized structure of the first excited triplet state T_1 at the TD-DFT level of theory.

triplet states	E_{rel}/eV	oscillator strength
T_2	2.08	0.127
T_3	2.37	0.643
T_4	2.75	$0.105 \cdot 10^{-6}$
T_5	3.12	0.210
T_6	3.25	$0.123 \cdot 10^{-3}$

Table B.23: Excitation energies of selected triplet states and corresponding oscillator strengths (vertical transitions) of **p-NP** relative to the optimized structure of the first excited triplet state T_1 at the TD-DFT level of theory.

triplet states	E_{rel}/eV	oscillator strength
T_2	1.30	0.277
T_3	1.31	$0.126 \cdot 10^{-7}$
T_4	2.59	0.799
T_5	3.43	$0.574 \cdot 10^{-3}$
T_6	3.64	0.251

B.3 Multireference methods

Table B.24: Cartesian coordinates in Å of the minimum energy conical intersection (MECI) S_0–S_1 geometry of *o*-**NP** optimized by CASSCF.

Atom	x	y	z
C	1.81858	-1.00076	-0.03172
C	1.8749	0.40617	0.22979
C	0.73376	1.15439	0.20418
C	-0.46699	0.52166	-0.10396
C	-0.59583	-0.90751	-0.30136
C	0.65984	-1.63997	-0.2883
H	2.75535	-1.56829	-0.01358
H	2.82183	0.86682	0.43856
H	0.75127	2.21231	0.38957
H	0.61603	-2.69591	-0.47713
N	-1.69775	1.20966	-0.1575
O	-1.70182	-1.41498	-0.40392
H	-2.84002	0.19772	0.86426
O	-2.52173	1.10404	0.89999
O	-2.20741	1.55463	-1.2489

B.4 Transient absorption spectroscopy

B.4.1 *ortho*-nitrophenol (*o*-NP)

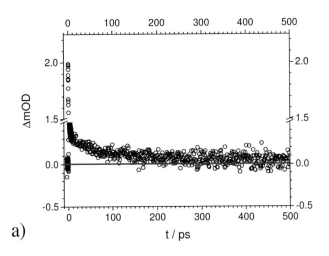

a)

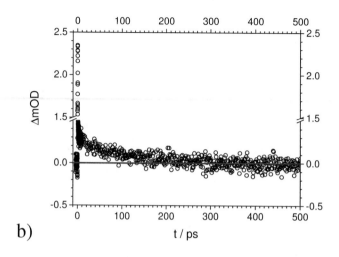

b)

Figure B.1: Transient absorption-time profile of *o*-**NP** in 2-propanol at a pump wavelength of 350 nm and a probe wavelength of **480 nm (a)** and **500 nm (b)**. A residual weak absorption is observable within the investigated time window of 500 ps, becoming progressively smaller for higher probe wavelengths.

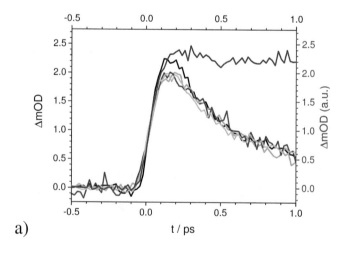

a)

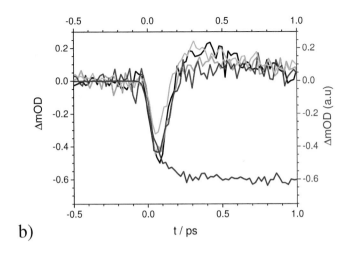

b)

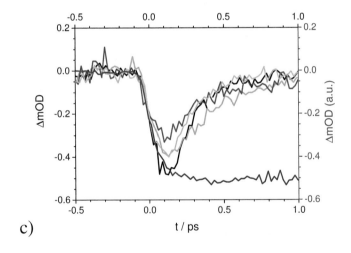

c)

Figure B.2: Comparison between reference experiments (blue; right axis) and TA traces of *o*-NP (left axis) in *n*-hexane (black), chloroform (green), 2-propanol (red) and water (cyan) at an excitation wavelength of 350 nm and the three selected probe wavelengths **480 nm (a)**, **600 nm (b)** and **900 nm (c)**. The reference experiments are cross correlation experiments of the dye BiBuQ (4,4'''-Bis-(2-butyloctyloxy)-p-quaterphenyl) in 1,4-dioxane for independent determination of time zero as well as the experimental time resolution (see section 3.5.1). For the sake of clarity, the TA profiles of the dye are scaled to 2.3, -0.6 and -0.5 ΔmOD at 480, 600 and 900 nm, respectively. Furthermore, they are inverted at 600 and 900 nm as the traces of *o*-NP reveal SE for these probe wavelengths. The quality of the experimental time resolution as well as the quality of time zero is demonstrated.

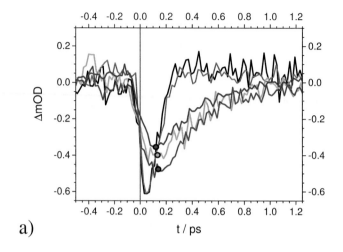

a)

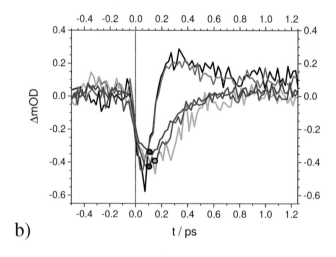

b)

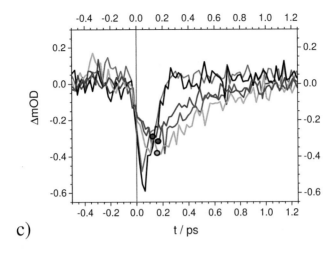

c)

Figure B.3: Selected TA profiles of *o*-**NP** in **chloroform (a)**, **2-propanol (b)** and **water (c)** at 650 nm (black), 700 nm (grey), 900 nm (blue), 950 nm (red) and 1100 nm (green). Only an indication of red-shifted SE is given (see circles to guide the eye).

B.4.2 *meta*- and *para*-nitrophenol (*m*- and *p*-NP)

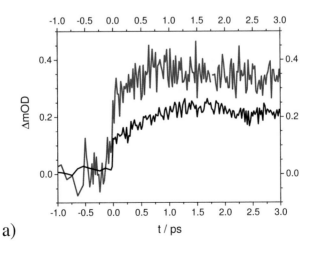

a)

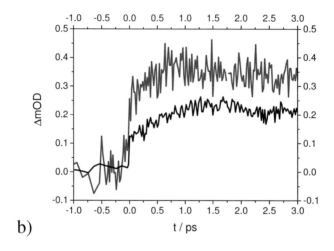

b)

Figure B.4: TA profiles of *p*-**NP** at an excitation wavelengths of 314 nm and the two probe wavelengths 900 nm (blue) and 950 nm (black) in **2-propanol (a)** and **chloroform (b)**. The traces are characterized by low intensity and in particular for 900 nm by a poor signal-to-noise ratio. Therefore they are not included in the global fit (see section 4.3.2.1).

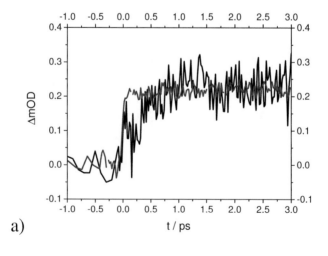

a)

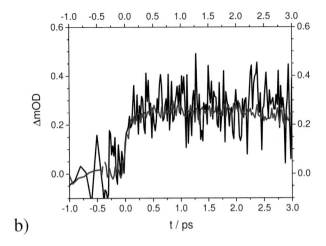

b)

t / ps

Figure B.5: TA profiles of **m-NP** at an excitation wavelengths of 314 nm and the two probe wavelengths 900 nm (blue) and 950 nm (black) in **2-propanol (a)** and **chloroform (b)**. The traces are characterized by low intensity and in particular for 950 nm by a poor signal-to-noise ratio. Therefore they are not included in the global fit (see section 4.3.2.2).

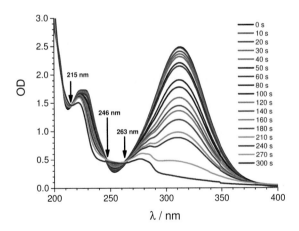

Figure B.6: Photolysis of **p-NP** in **2-propanol** performed in a static cuvette. The irradiation times are labeled in the graph. Three isosbestic points (marked by arrows) can be distinguished indicating photodegradation of the sample. An analog behavior was found for p-NP in chloroform, as well as m-NP in chloroform and 2-propanol (not shown here).

B.4.3 Nitrophenolates

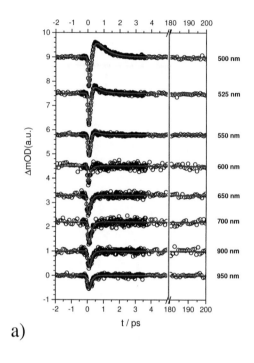

a)

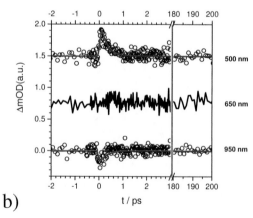

b)

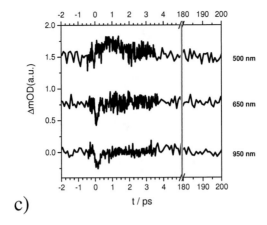

c)

Figure B.7: TA profiles of *p*-NP$^\ominus$ (**a**), *m*-NP$^\ominus$ (**b**) and *o*-NP$^\ominus$ (**c**) in **water** at a pump wavelength of 418 nm and probe wavelengths as labeled. Fit functions are displayed as red lines. TA profiles of *o*-NP$^\ominus$ could not be analyzed. Note that the plots are shifted along the ordinate for better clarity. For discussion see chapter 4, section 4.4.

C. Appendix for Chapter 5

C.1 Ground state geometry optimization

Table C.1: Cartesian coordinates in Å of the ground state minimum of **Bz** optimized by DFT.

Atom	x	y	z
C	-3.10817	-1.11129	0.34178
C	-1.70767	-1.87719	2.48727
C	-2.14613	-4.25295	3.58309
C	-3.98907	-5.87691	2.57207
C	-5.39754	-5.12602	0.44312
C	-4.95576	-2.76476	-0.66358
H	-0.28646	-0.60038	3.30008
H	-1.03833	-4.84110	5.24747
H	-4.33063	-7.73873	3.44541
H	-6.84239	-6.39797	-0.35591
H	-6.01582	-2.13687	-2.34241
C	-2.73146	1.36331	-0.99341
O	-3.87615	1.85790	-2.95334
C	-0.83731	3.37359	0.02273
O	-0.89626	3.69960	2.69044
H	-2.61680	4.19777	3.16728
H	-1.42176	5.15021	-0.95824
C	1.85214	2.73357	-0.80377
C	2.37017	2.22384	-3.37022
C	4.84232	1.67165	-4.15620
C	6.82935	1.62566	-2.39163
C	6.32321	2.14404	0.16176
C	3.84885	2.69675	0.95642
H	0.81907	2.26634	-4.76148
H	5.21933	1.27572	-6.16751
H	8.77137	1.19133	-3.01048
H	7.87144	2.12619	1.55771
H	3.45051	3.12667	2.95156

Table C.2: Cartesian coordinates in Å of the ground state minimum of **4MB** optimized by DFT.

Atom	x	y	z
C	1.06265	-2.43017	-0.23850
C	0.25551	-1.50994	2.14176
C	-1.24499	-3.01823	3.71649
C	-1.97507	-5.48602	3.00001
C	-1.15444	-6.39466	0.62618
C	0.33037	-4.89589	-0.96655
H	0.81743	0.39159	2.75731
H	-1.86804	-2.26260	5.55818
C	-3.58610	-7.08732	4.73140
H	-1.70143	-8.31429	0.02182
H	0.95624	-5.58972	-2.82687
C	2.61492	-0.94404	-2.08233
O	3.07516	-1.76473	-4.20796
C	3.66792	1.71744	-1.39549
O	4.59915	1.92896	1.11749
H	5.92069	0.64960	1.34559
H	5.22289	2.01485	-2.79398
C	1.66392	3.74463	-1.84375
C	1.09388	5.53874	0.03850
C	-0.72052	7.42593	-0.41612
C	-1.98042	7.54279	-2.74951
C	-1.41018	5.75974	-4.63642
C	0.40077	3.87402	-4.18963
H	2.10133	5.45694	1.85541
H	-1.14546	8.82340	1.07132
H	-3.40259	9.02458	-3.10274
H	-2.38187	5.83815	-6.47862
H	0.84911	2.48400	-5.67602
H	-3.99047	-8.97058	3.89791
H	-2.64159	-7.39315	6.58769
H	-5.42874	-6.15401	5.13740

Table C.3: Cartesian coordinates in Å of the ground state minimum of **MMMP** optimized by DFT.

Atom	x	y	z
C	3.10053	-2.09928	-0.42391
C	3.29081	-2.89889	2.36819
C	5.70056	-2.58657	-1.64583
N	1.12425	-3.70616	-1.66105
C	2.31820	0.70611	-1.01143
O	2.62137	1.45426	-3.19932
C	0.87959	2.38174	0.78715
C	1.29451	2.58639	3.42414
C	-0.04404	4.33420	4.88891
C	-1.88680	5.92593	3.77961
C	-2.30424	5.75436	1.14435
C	-0.91430	4.03800	-0.30772
H	2.74342	1.43588	4.36384
H	0.34937	4.48042	6.93078
S	-3.49075	8.01605	5.84472
H	-3.70460	6.96999	0.20391
H	-1.19603	3.94671	-2.36842
C	-5.59763	9.77686	3.77380
H	-6.61531	11.11509	5.02830
H	-4.53361	10.87376	2.33455
H	-7.00160	8.52434	2.84201
H	5.81062	-1.77777	-3.57428
H	7.19952	-1.66602	-0.49696
H	6.09151	-4.65023	-1.71444
H	1.54606	-2.55687	3.47547
H	3.69714	-4.96011	2.43355
H	4.86732	-1.89586	3.32618
C	1.06277	-3.82229	-4.42544
C	-0.57776	-6.06335	-5.26673
H	0.29912	-2.03723	-5.27679
H	2.99538	-4.09963	-5.18296
O	-3.07407	-5.91054	-4.29921
H	-0.74528	-6.08665	-7.35973
H	0.33920	-7.86778	-4.63777
C	-3.02589	-5.77132	-1.62471
C	-1.47118	-3.50476	-0.70252
H	-5.02084	-5.59038	-0.99302
H	-2.20976	-7.54837	-0.80779
H	-2.43691	-1.73318	-1.36282
H	-1.48062	-3.48681	1.39341

C.2 Excitation energies related to the ground state

Table C.4: Excitation energies of selected singlet and triplet states as well as oscillator strengths of vertical transitions of **Bz** relative to the optimized ground state structure at the TD-DFT level of theory.

Singlet states	E_{rel}/eV	Oscillator strength	Triplet states	E_{rel}/eV
S_1	3.41	$0.661 \cdot 10^{-2}$	T_1	2.97
S_2	4.20	$0.330 \cdot 10^{-1}$	T_2	3.17
S_3	4.27	$0.074 \cdot 10^{-2}$	T_3	3.59
S_4	4.44	$0.183 \cdot 10^{-1}$	T_4	3.73
S_5	4.85	0.250	T_5	4.10

Table C.5: Excitation energies of selected singlet and triplet states as well as oscillator strengths of vertical transitions of **4MB** relative to the optimized ground state structure at the TD-DFT level of theory.

Singlet states	E_{rel}/eV	Oscillator strength	Triplet states	E_{rel}/eV
S_1	3.44	$0.673 \cdot 10^{-2}$	T_1	3.00
S_2	4.24	$0.604 \cdot 10^{-1}$	T_2	3.08
S_3	4.32	$0.116 \cdot 10^{-2}$	T_3	3.59
S_4	4.45	$0.187 \cdot 10^{-1}$	T_4	3.79
S_5	4.68	0.299	T_5	4.15

Table C.6: Excitation energies of selected singlet and triplet states as well as oscillator strengths of vertical transitions of **MMMP** relative to the optimized ground state structure at the TD-DFT level of theory.

Singlet states	E_{rel}/eV	Oscillator strength	Triplet states	E_{rel}/eV
S_1	3.14	$0.311 \cdot 10^{-1}$	T_1	2.73
S_2	3.98	0.339	T_2	2.87
S_3	4.23	$0.575 \cdot 10^{-1}$	T_3	3.82
S_4	4.36	$0.460 \cdot 10^{-3}$	T_4	3.94
S_5	4.45	$0.171 \cdot 10^{-1}$	T_5	4.16

C.3 Transient absorption spectroscopy

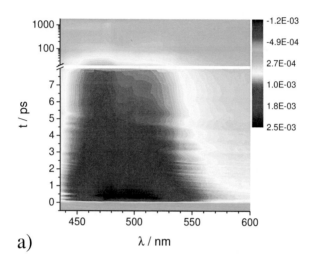

a)

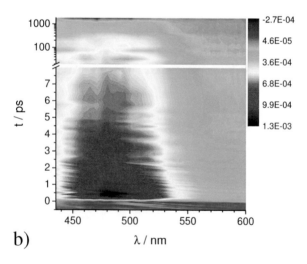

b)

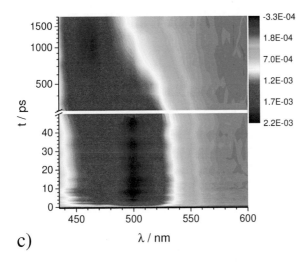

c)

Figure C.1: TA profiles of **Bz (a)**, **4MB (b)** and **MMMP (c)** in **MIB** at a pump wavelength of 325 nm and probed using a white light continuum between 435 and 600 nm.

Table C.7: Time constants of **Bz**, **4MB** and **MMMP** obtained by global fit routine in MIB at 325 nm. The given errors are standard deviations derived from the fitting procedure. For τ_2 (Bz and 4MB) and τ_3 (MMMP) a range of time constants is given. The obtained values are in good agreement with those derived at 351 nm (chapter 5, section 5.3).

photoinitiator	τ_1 / ps	τ_2 / ps	τ_3 / ps	τ_4 / ps
Bz	2.5 ± 0.1	(50-80)	>1800	---
4MB	2.5 ± 0.1	(150-350)	>1800	---
MMMP	1.1 ± 0.1	57 ± 6	(450-650)	>1800

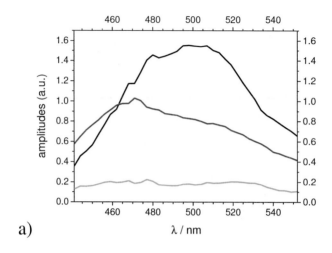

a)

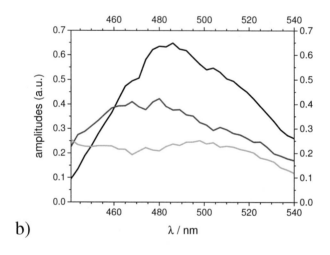

b)

Figure C.2: DADS spectra for **Bz (a)** and **4MB (b)** in **MIB** at a pump wavelength of 325 nm. Amplitude A is displayed by black, amplitude B by red and amplitude C by green lines. Analogously to the DADS spectra obtained at 351 nm, a similar trend for the amplitudes is observed for Bz and 4MB.

Table C.8: Relative amplitudes of **Bz** and **4MB** at a pump wavelength of 325 nm and three selected probe wavelengths as labeled. The obtained values at 325 nm are in good accordance to those derived at 351 nm (chapter 5, section 5.4.1).

photoinitiator		A_{rel}	B_{rel}	C_{rel}
Bz	477 nm	0.54	0.37	0.09
	489 nm	0.58	0.35	0.07
	501 nm	0.61	0.32	0.07
4MB	477 nm	0.50	0.33	0.17
	489 nm	0.50	0.30	0.20
	501 nm	0.49	0.29	0.22

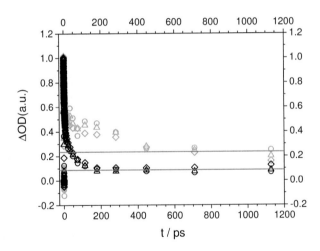

Figure C.3: Normalized TA traces of **Bz (black)** and **4MB (green)** in **MIB** at the three selected probe wavelengths 477 nm (squares), 489 nm (triangles) and 501 nm (circles) and at a pump wavelength of 325 nm. ΔOD (a.u.) values of 0.09 for Bz and 0.23 for 4MB are indicated by grey lines. The relation between 4MB and Bz is determined by a factor of ~2.5. Thus, a higher ability of radical formation is found for 4MB in comparison to Bz in agreement with the value of ~3.4 derived at 351 nm (chapter 5, section 5.4.1).

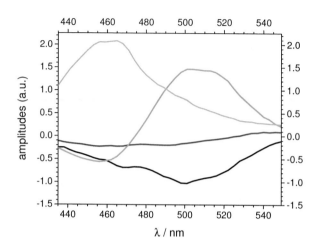

Figure C.4: DADS spectrum for **MMMP** in **MIB** at a pump wavelength of 325 nm. Amplitude A is displayed by black, amplitude B by red, amplitude C by green and amplitude D by cyan lines. Compared to the DADS spectrum at 351 nm an analog trend for the amplitudes is observed (chapter 5, section 5.4.2).

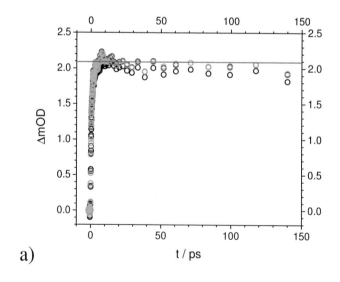

a)

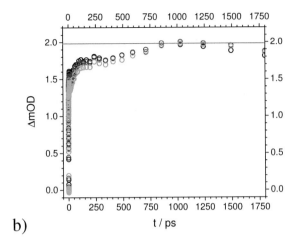

b)

Figure C.5: Singlet absorption band of **MMMP (a)** at 525 nm (black), 522 nm (red) and 519 nm (green) as well as **triplet absorption band (b)** at 471 nm (black), 468 nm (red) and 465 nm (green). The pump wavelength was at 325 nm. ΔOD (a.u.) values of 2.09 (a) and 1.98 (b) are indicated by grey lines, respectively. The relation between the obtained values is ~0.95 which indicates a population of T_1 by 95% assuming comparable absorption properties of both excited species. These findings agree very well with those obtained from 351 nm (chapter 5, section 5.4.2).

References

1.   Zewail, A. H. *J. Phys. Chem. A* **2000,** 104, (24), 5660-5694.

2.   Maiman, T. H. *Br. Commun. Electron.* **1960,** 7, 674-5.

3.   Hellwarth, R. W. *Lasers* **1966,** 1, 253-94.

4.   DeMaria, A. J.; Stetser, D. A.; Heynau, H. *Appl. Phys. Lett.* **1966,** 8, (7), 1746.

5.   Stetser, D. A.; DeMaria, A. J. *Appl. Phys. Lett.* **1966,** 9, (3), 118-20.

6.   Fork, R. L.; Cruz, C. H. B.; Becker, P. C.; Shank, C. V. *Opt. Lett.* **1987,** 12, (7), 483-5.

7.   Spence, D. E.; Kean, P. N.; Sibbett, W. *Opt. Lett.* **1991,** 16, (1), 42-4.

8.   Zhou, J.; Taft, G.; Huang, C.-P.; Murnane, M. M.; Kapteyn, H. C. *Opt. Lett.* **1994,** 19, (15), 1149-51.

9.   Baltuska, A.; Wei, Z.; Pshenichnikov, M. S.; Wiersma, D. A. *Opt Lett* **1997,** 22, (2), 102-4.

10.   Kryukov, P. G. *Quantum Electron.* **2001,** 31, (2), 95-119.

11.   French, P. M. W. *Rep. Prog. Phys.* **1995,** 58, (2), 169-262.

12.   Morgner, U.; Kartner, F. X.; Cho, S. H.; Chen, Y.; Haus, H. A.; Fujimoto, J. G.; Ippen, E. P.; Scheuer, V.; Angelow, G.; Tschudi, T. *Opt Lett* **1999,** 24, (6), 411-3.

13.   Alfano, R. R.; Shapiro, S. L. *Phys. Rev. Lett.* **1970,** 24, (11), 592-4.

14.   Piel, J.; Beutter, M.; Riedle, E. *Opt Lett* **2000,** 25, (3), 180-2.

15.   Wilhelm, T.; Piel, J.; Riedle, E. *Opt Lett* **1997,** 22, (19), 1494-6.

16.   Pedersen, S.; Zewail, A. H. *Mol. Phys.* **1996,** 89, (5), 1455-1502.

17.   Polli, D.; Altoe, P.; Weingart, O.; Spillane, K. M.; Manzoni, C.; Brida, D.; Tomasello, G.; Orlandi, G.; Kukura, P.; Mathies, R. A.; Garavelli, M.; Cerullo, G. *Nature (London, U. K.)* **2010,** 467, (7314), 440-443.

18.   Yoshizawa, T.; Wald, G. *Nature (London, U. K.)* **1963,** 197, 1279-86.

19.   Hecht, S.; Shlaer, S.; Pirenne, M. H. *J Gen Physiol* **1942,** 25, (6), 819-40.

20.   Kochendoerfer, G. G.; Mathies, R. A. *J. Phys. Chem.* **1996,** 100, (34), 14526-14532.

21. Schoenlein, R. W.; Peteanu, L. A.; Mathies, R. A.; Shank, C. V. *Science (Washington, D. C., 1883-)* **1991,** 254, (5030), 412-15.

22. Haran, G.; Morlino, E. A.; Matthes, J.; Callender, R. H.; Hochstrasser, R. M. *J. Phys. Chem. A* **1999,** 103, (14), 2202-2207.

23. Chosrowjan, H.; Mataga, N.; Shibata, Y.; Tachibanaki, S.; Kandori, H.; Shichida, Y.; Okada, T.; Kouyama, T. *J. Am. Chem. Soc.* **1998,** 120, (37), 9706-9707.

24. Kandori, H.; Furutani, Y.; Nishimura, S.; Shichida, Y.; Chosrowjan, H.; Shibata, Y.; Mataga, N. *Chem. Phys. Lett.* **2001,** 334, (4,5,6), 271-276.

25. Kim, J. E.; Tauber, M. J.; Mathies, R. A. *Biochemistry* **2001,** 40, (46), 13774-13778.

26. Cooper, A. *Nature* **1979,** 282, (5738), 531-3.

27. Garavelli, M.; Celani, P.; Bernardi, F.; Robb, M. A.; Olivucci, M. *J. Am. Chem. Soc.* **1997,** 119, (29), 6891-6901.

28. Abe, M.; Ohtsuki, Y.; Fujimura, Y.; Domcke, W. *J. Chem. Phys.* **2005,** 123, (14), 144508/1-144508/10.

29. P. Atkins, R. F. *Molecular Quantum Mechanics, 4th edition, Oxford University Press Inc., New York, 2005.*

30. C. N. Banwell, E. M. M. *Fundamentals of Molecular Spectroscopy,* 4th edition, McGraw-Hill Book Company, UK, **1994**.

31. Jablonski, A. *Z. Phys.* **1935,** 94, 38-46.

32. Kasha, M. *Discuss. Faraday Soc.* **1950,** No. 9, 14-19.

33. Coto, P. B.; Serrano-Andres, L.; Gustavsson, T.; Fujiwara, T.; Lim, E. C. *Phys Chem Chem Phys* **2011,** 13, (33), 15182-8.

34. Thomsen, C. L.; Thogersen, J.; Keiding, S. R. *J. Phys. Chem. A* **1998,** 102, (7), 1062-1067.

35. Grabowski, Z. R.; Rotkiewicz, K.; Rettig, W. *Chem. Rev. (Washington, DC, U. S.)* **2003,** 103, (10), 3899-4031.

36. Poor, B.; Michniewicz, N.; Kallay, M.; Buma, W. J.; Kubinyi, M.; Szemik-Hojniak, A.; Deperasinska, I.; Puszko, A.; Zhang, H. *J. Phys. Chem. A* **2006,** 110, (22), 7086-7091.

37. Druzhinin, S. I.; Mayer, P.; Stalke, D.; von Buelow, R.; Noltemeyer, M.; Zachariasse, K. A. *J. Am. Chem. Soc.* **2010,** 132, (22), 7730-7744.

38. Zilberg, S.; Haas, Y. *J. Phys. Chem. A* **2002,** 106, (1), 1-11.

39. Cogan, S.; Zilberg, S.; Haas, Y. *J. Am. Chem. Soc.* **2006,** 128, (10), 3335-3345.

40. Atsbeha, T.; Mohammed, A. M.; Redi-Abshiro, M. *J. Fluoresc.* **2010,** 20, (6), 1241-1248.

41. Grabowski, Z. R.; Rotkiewicz, K.; Siemiarczuk, A. *J. Lumin.* **1979,** 18-19, (Pt. 1), 420-4.

42. Rotkiewicz, K.; Grellmann, K. H.; Grabowski, Z. R. *Chem. Phys. Lett.* **1973,** 19, (3), 315-18.

43. de Klerk, J. S.; Szemik-Hojniak, A.; Ariese, F.; Gooijer, C. *J. Phys. Chem. A* **2007,** 111, (26), 5828-5832.

44. Hecht, E. *Optics,* 4th edition, Pearson, San Francisco, USA, **2002.**

45. J.-C. Diels, W. R. *Ultrashort Laser Pulse Phenomena,* 2nd edition, Academic Press, San Diego, USA, **2006.**

46. Fieß, M. *Aufbau eines FROG-Systems zur Charakterisierung von Femtosekunden-Laserpulsen,* Diploma thesis, University Karlsruhe, **2006.**

47. Rulliere, C. *Femtosecond Laser Pulses, Principles and Experiments,* 2nd edition, Springer, New York, USA, **2005.**

48. Boyd, R. W. *Nonlinear optics,* 3rd edition, Academic Press, Burlington, USA, **2008.**

49. Franken, P. A.; Hill, A. E.; Peters, C. W.; Weinreich, G. *Physical Review Letters* **1961,** 7, (4), 118-119.

50. *USER`S MANUAL for CPA-2001,* Clark-MXR Inc., Version 4.1, **2002.**

51. Brands, H. *Ultrakurzzeitdynamik von Fulleriden in Lösung und suspendierten, längenselektierten Kohlenstoffnanoröhren,* Dissertation, University Karlsruhe (TH), **2007,** ISBN-13: 978-3866442115.

52. Tamura, K.; Doerr, C. R.; Nelson, L. E.; Haus, H. A.; Ippen, E. P. *Opt. Lett.* **1994,** 19, (1), 46-8.

53. Haus, H. A.; Ippen, E. P.; Tamura, K. *IEEE J. Quantum Electron.* **1994,** 30, (1), 200-8.

54. Haus, H.; Fujimoto, J. G.; Ippen, E. P. *Quantum Electronics, IEEE Journal of* **1992,** 28, (10), 2086-2096.

55. Takada, H.; Kakehata, M.; Torizuka, K. *Opt. Lett.* **2006,** 31, (8), 1145-1147.

56. Maine, P.; Strickland, D.; Bado, P.; Pessot, M.; Mourou, G. *IEEE J. Quantum Electron.* **1988,** QE-24, (2), 398-403.

57. Strickland, D.; Mourou, G. *Opt. Commun.* **1985,** 55, (6), 447-9.

58. Riedle, E.; Beutter, M.; Lochbrunner, S.; Piel, J.; Schenkl, S.; Sporlein, S.; Zinth, W. *Appl. Phys. B: Lasers Opt.* **2000,** 71, (3), 457-465.

59. Megerle, U.; Pugliesi, I.; Schriever, C.; Sailer, C. F.; Riedle, E. *Appl. Phys. B: Lasers Opt.* **2009,** 96, (2-3), 215-231.

60. Tarnovsky, A. N.; Gawelda, W.; Johnson, M.; Bressler, C.; Chergui, M. *J. Phys. Chem. B* **2006,** 110, (51), 26497-26505.

61. Zamyatin, A. V.; Soldatova, A. V.; Rodgers, M. A. J. *Inorg. Chim. Acta* **2007,** 360, (3), 857-868.

62. Ernsting, N. P.; Kovalenko, S. A.; Senyushkina, T.; Saam, J.; Farztdinov, V. *J. Phys. Chem. A* **2001,** 105, (14), 3443-3453.

63. Bizjak, T.; Karpiuk, J.; Lochbrunner, S.; Riedle, E. *J. Phys. Chem. A* **2004,** 108, (49), 10763-10769.

64. Schmidt, B.; Sobotta, C.; Malkmus, S.; Laimgruber, S.; Braun, M.; Zinth, W.; Gilch, P. *J. Phys. Chem. A* **2004,** 108, (20), 4399-4404.

65. Karunakaran, V.; Pfaffe, M.; Ioffe, I.; Senyushkina, T.; Kovalenko, S. A.; Mahrwald, R.; Fartzdinov, V.; Sklenar, H.; Ernsting, N. P. *J. Phys. Chem. A* **2008,** 112, (18), 4294-4307.

66. Laimgruber, S.; Schachenmayr, H.; Schmidt, B.; Zinth, W.; Gilch, P. *Appl. Phys. B: Lasers Opt.* **2006,** 85, (4), 557-564.

67. Unterreiner, A.-N. *Ultraschnelle Relaxationsdynamik solvatisierter Elektronen in flüssigem Ammoniak und Wasser,* Dissertation, University Karlsruhe (TH), **1998,** ISBN: 3-89653-368-1.

68. Liang, Y.; Bradler, M.; Klinger, M.; Schalk, O.; Balaban, M. C.; Balaban, T. S.; Riedle, E.; Unterreiner, A.-N. *ChemPlusChem* **2013,** 78, (10), 1244-1251.

69. Bradler, M.; Baum, P.; Riedle, E. *Appl. Phys. B: Lasers Opt.* **2009,** 97, (3), 561-574.

70. Jensen, F. *Introduction to Computational Chemistry,* 2nd edition, Wiley, West-Sussex, England, **2007.**

71. Cramer, C. J. *Essentials of Computational Chemistry, Theories and Models,* 2nd edition, Wiley, West-Sussex, England, **2004.**

72. Hohenberg, P.; Kohn, W. *Physical Review* **1964,** 136, (3B), B864-B871.

73. Kohn, W.; Sham, L. J. *Physical Review* **1965,** 140, (4A), A1133-A1138.

74. Vosko, S. H.; Wilk, L.; Nusair, M. *Can. J. Phys.* **1980,** 58, (8), 1200-11.

75. Becke, A. D. *Phys. Rev. A: Gen. Phys.* **1988,** 38, (6), 3098-100.

76. Becke, A. D. *J. Chem. Phys.* **1993,** 98, (7), 5648-52.

77. Stephens, P. J.; Devlin, F. J.; Chabalowski, C. F.; Frisch, M. J. *J. Phys. Chem.* **1994,** 98, (45), 11623-7.

78. Runge, E.; Gross, E. K. U. *Phys. Rev. Lett.* **1984,** 52, (12), 997-1000.

79. Elliott, P.; Furche, F.; Burke, K., Excited States from Time-Dependent Density Functional Theory. In *Reviews in Computational Chemistry,* John Wiley & Sons, Inc.: 2009; pp 91-165.

80. Levine, B. G.; Ko, C.; Quenneville, J.; Martinez, T. J. *Mol. Phys.* **2006,** 104, (5-7), 1039-1051.

81. P. W. Atkins, J. d. P. *Physikalische Chemie,* 4. Aufl. Wiley-VCH, Weinheim, Germany, **2006.**

82. Kozma, I.; Baum, P.; Lochbrunner, S.; Riedle, E. *Opt Express* **2003,** 11, (23), 3110-5.

83. Diels, J. C. M.; Fontaine, J. J.; McMichael, I. C.; Simoni, F. *Appl. Opt.* **1985,** 24, (9), 1270-82.

84. Maciejewski, A.; Naskrecki, R.; Lorenc, M.; Ziolek, M.; Karolczak, J.; Kubicki, J.; Matysiak, M.; Szymanski, M. *J. Mol. Struct.* **2000,** 555, 1-13.

85. Raytchev, M.; Pandurski, E.; Buchvarov, I.; Modrakowski, C.; Fiebig, T. *J. Phys. Chem. A* **2003,** 107, (23), 4592-4600.

86. Klimov, V. I.; McBranch, D. W. *Opt. Lett.* **1998,** 23, (4), 277-279.

87. Kovalenko, S. A.; Ernsting, N. P.; Ruthmann, J. *Chem. Phys. Lett.* **1996,** 258, (3,4), 445-454.

88. Brackmann, U., *LambdachromeLaser®Dyes,* Goettingen, Germany, **2000.**

89. Wolf, T. J. A.; Fischer, J.; Wegener, M.; Unterreiner, A.-N. *Opt. Lett.* **2011,** 36, (16), 3188-3190.

90. Fischer, J.; Ergin, T.; Wegener, M. *Opt Lett* **2011,** 36, (11), 2059-61.

91. Voll, D.; Hufendiek, A.; Junkers, T.; Barner-Kowollik, C. *Macromol. Rapid Commun.* **2012,** 33, (1), 47-53.

92. Ahlrichs, R.; Baer, M.; Haeser, M.; Horn, H.; Koelmel, C. *Chem. Phys. Lett.* **1989,** 162, (3), 165-9.

93. Dirac, P. A. M. *Proc. R. Soc. London, Ser. A* **1929,** 123, 714-33.

94. Eichkorn, K.; Treutler, O.; Oehm, H.; Haeser, M.; Ahlrichs, R. *Chem. Phys. Lett.* **1995,** 240, (4), 283-90.

95. Eichkorn, K.; Treutler, O.; Oehm, H.; Haeser, M.; Ahlrichs, R. *Chem. Phys. Lett.* **1995,** 242, (6), 652-60.

96. Eichkorn, K.; Weigend, F.; Treutler, O.; Ahlrichs, R. *Theor. Chem. Acc.* **1997,** 97, (1-4), 119-124.

97. Schaefer, A.; Horn, H.; Ahlrichs, R. *J. Chem. Phys.* **1992,** 97, (4), 2571-7.

98.   Slater, J. C. *Phys. Rev.* **1951,** 81, 385-90.

99.   Treutler, O.; Ahlrichs, R. *J. Chem. Phys.* **1995,** 102, (1), 346-54.

100. Weigend, F. *Phys. Chem. Chem. Phys.* **2006,** 8, (9), 1057-1065.

101. *USER`S MANUAL for Turbomole Version 6.3,* **2011.**

102. Bauernschmitt, R.; Ahlrichs, R. *J. Chem. Phys.* **1996,** 104, (22), 9047-9052.

103. Bauernschmitt, R.; Ahlrichs, R. *Chem. Phys. Lett.* **1996,** 256, (4,5), 454-464.

104. Grimme, S.; Furche, F.; Ahlrichs, R. *Chem. Phys. Lett.* **2002,** 361, (3,4), 321-328.

105. Lee, C.; Yang, W.; Parr, R. G. *Phys. Rev. B: Condens. Matter* **1988,** 37, (2), 785-9.

106. Kendall, R. A.; Dunning, T. H., Jr.; Harrison, R. J. *J. Chem. Phys.* **1992,** 96, (9), 6796-806.

107. Schmidt, M. W.; Baldridge, K. K.; Boatz, J. A.; Elbert, S. T.; Gordon, M. S.; Jensen, J. H.; Koseki, S.; Matsunaga, N.; Nguyen, K. A.; et, a. *J. Comput. Chem.* **1993,** 14, (11), 1347-63.

108. Gordon, M. S.; Schmidt, M. W. In *Advances in electronic structure theory: GAMESS a decade later,* **2005;** Elsevier B.V.: 2005; pp 1167-1189.

109. Ditchfield, R.; Hehre, W. J.; Pople, J. A. *J. Chem. Phys.* **1971,** 54, (2), 724-8.

110. Hariharan, P. C.; Pople, J. A. *Theor. Chim. Acta* **1973,** 28, (3), 213-22.

111. Hariharan, P. C.; Pople, J. A. *Mol. Phys.* **1974,** 27, (1), 209-14.

112. Hehre, W. J.; Ditchfield, R.; Pople, J. A. *J. Chem. Phys.* **1972,** 56, (5), 2257-61.

113. Nakano, H. *J. Chem. Phys.* **1993,** 99, (10), 7983-92.

114. Nakano, H. *Chem. Phys. Lett.* **1993,** 207, (4-6), 372-8.

115. Aquilante, F.; De Vico, L.; Ferre, N.; Ghigo, G.; Malmqvist, P.-a.; Neogrady, P.; Pedersen, T. B.; Pitonak, M.; Reiher, M.; Roos, B. O.; Serrano-Andres, L.; Urban, M.; Veryazov, V.; Lindh, R. *J. Comput. Chem.* **2010,** 31, (1), 224-247.

116. Veryazov, V.; Widmark, P.-O.; Serrano-Andres, L.; Lindh, R.; Roos, B. O. *Int. J. Quantum Chem.* **2004,** 100, (4), 626-635.

117. Karlstroem, G.; Lindh, R.; Malmqvist, P.-A.; Roos, B. O.; Ryde, U.; Veryazov, V.; Widmark, P.-O.; Cossi, M.; Schimmelpfennig, B.; Neogrady, P.; Seijo, L. *Comput. Mater. Sci.* **2003,** 28, (2), 222-239.

118. Frisch, M. J. T., G. W.;Schlegel, H. B.; Scuseria, G. E.; Robb, M. A.; Cheeseman, J. R.; Scalmani, G.; Barone, V.; et al. . **2009.**

119. H. A. Ernst, T. J. A. W., O. Schalk, N. Gonzalez-Garcia, A. E. Boguslavskiy, A. Stolow, M. Olzmann, A.-N. Unterreiner. *to be submitted* **2015.**

120. Haynes, W. M., *Handbook of Chemisty and Physics*, Aufl. 91, CRC press, London, **2011**.

121. Atkinson, R. *Atmos. Environ.* **2000,** 34, (12-14), 2063-2101.

122. Harrison, M. A. J.; Barra, S.; Borghesi, D.; Vione, D.; Arsene, C.; Olariu, R. I. *Atmos. Environ.* **2005,** 39, (2), 231-248.

123. Tremp, J.; Mattrel, P.; Fingler, S.; Giger, W. *Water, Air, Soil Pollut.* **1993,** 68, (1-2), 113-23.

124. Bejan, I.; Abd El Aal, Y.; Barnes, I.; Benter, T.; Bohn, B.; Wiesen, P.; Kleffmann, J. *Phys. Chem. Chem. Phys.* **2006,** 8, (17), 2028-2035.

125. Derwent, R. G.; Jenkin, M. E.; Saunders, S. M. *Atmos. Environ.* **1996,** 30, (2), 181-99.

126. Derwent, R. G.; Jenkin, M. E.; Saunders, S. M.; Pilling, M. J. *Atmos. Environ.* **1998,** 32, (14/15), 2429-2441.

127. Harrison, R. M.; Peak, J. D.; Collins, G. M. *J. Geophys. Res., [Atmos.]* **1996,** 101, (D9), 14429-14439.

128. Alicke, B.; Platt, U.; Stutz, J. *J. Geophys. Res., [Atmos.]* **2002,** 107, (D22), LOP9/1-LOP9/17.

129. Perner, D.; Ehhalt, D. H.; Paetz, H. W.; Platt, U.; Roeth, E. P.; Volz, A. *Geophys. Res. Lett.* **1976,** 3, (8), 466-8.

130. Kurtenbach, R.; Becker, K. H.; Gomes, J. A. G.; Kleffmann, J.; Lorzer, J. C.; Spittler, M.; Wiesen, P.; Ackermann, R.; Geyer, A.; Platt, U. *Atmos. Environ.* **2001,** 35, (20), 3385-3394.

131. Finlayson-Pitts, B. J.; Wingen, L. M.; Sumner, A. L.; Syomin, D.; Ramazan, K. A. *Phys. Chem. Chem. Phys.* **2003,** 5, (2), 223-242.

132. Kleffmann, J.; Loerzer, J. C.; Wiesen, P.; Kern, C.; Trick, S.; Volkamer, R.; Rodenas, M.; Wirtz, K. *Atmos. Environ.* **2006,** 40, (20), 3640-3652.

133. Ramazan, K. A.; Syomin, D.; Finlayson-Pitts, B. J. *Phys. Chem. Chem. Phys.* **2004,** 6, (14), 3836-3843.

134. Stemmler, K.; Ammann, M.; Donders, C.; Kleffmann, J.; George, C. *Nature (London, U. K.)* **2006,** 440, (7081), 195-198.

135. Alif, A.; Pilichowski, J. F.; Boule, P. *J. Photochem. Photobiol., A* **1991,** 59, (2), 209-19.

136. Cheng, S.-B.; Zhou, C.-H.; Yin, H.-M.; Sun, J.-L.; Han, K.-L. *J. Chem. Phys.* **2009,** 130, (23), 234311/1-234311/8.

137. Nagaya, M.; Kudoh, S.; Nakata, M. *Chem. Phys. Lett.* **2006,** 427, (1-3), 67-71.

138. Wang, Y.-Q.; Wang, H.-G.; Zhang, S.-Q.; Pei, K.-M.; Zheng, X.; Phillips, D. L. *J. Chem. Phys.* **2006,** 125, (21), 214506/1-214506/12.

139. Alif, A.; Boule, P.; Lemaire, J. *Chemosphere* **1987,** 16, (10-12), 2213-23.

140. Alif, A.; Boule, P.; Lemaire, J. *J. Photochem. Photobiol., A* **1990,** 50, (3), 331-42.

141. Takezaki, M.; Hirota, N.; Terazima, M. *J. Phys. Chem. A* **1997,** 101, (19), 3443-3448.

142. Magdalena Szostak, M.; Kozankiewicz, B.; Wojcik, G. y.; Lipinski, J. *J. Chem. Soc., Faraday Trans.* **1998,** 94, (21), 3241-3245.

143. Carroll, M. K.; Hieftje, G. M. *Appl. Spectrosc.* **1992,** 46, (1), 126-30.

144. Namboodiri, K. P. K.; Viswanathan, S.; Ganesan, R.; Bhasu, V. C. J. *J. Comput. Chem.* **1981,** 2, (4), 392-401.

145. Nielsen, S. B.; Nielsen, M. B.; Rubio, A. *Acc. Chem. Res.* **2014,** 47, (4), 1417-1425.

146. Leavell, S.; Curl, R. F., Jr. *J. Mol. Spectrosc.* **1973,** 45, (3), 428-42.

147. Borisenko, K. B.; Bock, C. W.; Hargittai, I. *J. Phys. Chem.* **1994,** 98, (5), 1442-8.

148. Heinz, B.; Schmierer, T.; Laimgruber, S.; Gilch, P. *J. Photochem. Photobiol., A* **2008,** 199, (2-3), 274-281.

149. Farztdinov, V. M.; Schanz, R.; Kovalenko, S. A.; Ernsting, N. P. *J. Phys. Chem. A* **2000,** 104, (49), 11486-11496.

150. Kovalenko, S. A.; Schanz, R.; Farztdinov, V. M.; Hennig, H.; Ernsting, N. P. *Chem. Phys. Lett.* **2000,** 323, (3,4), 312-322.

151. Schriever, C.; Lochbrunner, S.; Ofial, A. R.; Riedle, E. *Chem. Phys. Lett.* **2011,** 503, (1-3), 61-65.

152. Barbatti, M.; Aquino, A. J. A.; Lischka, H.; Schriever, C.; Lochbrunner, S.; Riedle, E. *Phys. Chem. Chem. Phys.* **2009,** 11, (9), 1406-1415.

153. Lochbrunner, S.; Wurzer, A. J.; Riedle, E. *J. Phys. Chem. A* **2003,** 107, (49), 10580-10590.

154. Huston, A. L.; Scott, G. W.; Gupta, A. *J. Chem. Phys.* **1982,** 76, (10), 4978-85.

155. Hu, S.; Liu, K.; Li, Y.; Ding, Q.; Peng, W.; Chen, M. *Can. J. Chem.* **2014,** 92, (4), 274-278.

156. El-Sayed, M. A. *J. Chem. Phys.* **1963,** 38, 2834-8.

157. Lower, S. K.; El-Sayed, M. A. *Chem. Rev.* **1966,** 66, (2), 199-241.

158. Zugazagoitia, J. S.; Almora-Diaz, C. X.; Peon, J. *J. Phys. Chem. A* **2008,** 112, (3), 358-365.

159. Minns, R. S.; Parker, D. S. N.; Penfold, T. J.; Worth, G. A.; Fielding, H. H. *Phys. Chem. Chem. Phys.* **2010,** 12, (48), 15607-15615.

160. Wanko, M.; Houmoller, J.; Stochkel, K.; Suhr Kirketerp, M.-B.; Petersen, M. A.; Nielsen, M. B.; Nielsen, S. B.; Rubio, A. *Phys. Chem. Chem. Phys.* **2012**, 14, (37), 12905-12911.

161. Morales-Cueto, R.; Esquivelzeta-Rabell, M.; Saucedo-Zugazagoitia, J.; Peon, J. *J. Phys. Chem. A* **2007**, 111, (4), 552-557.

162. Collado-Fregoso, E.; Zugazagoitia, J. S.; Plaza-Medina, E. F.; Peon, J. *J. Phys. Chem. A* **2009**, 113, (48), 13498-13508.

163. Plaza-Medina, E. F.; Rodriguez-Cordoba, W.; Morales-Cueto, R.; Peon, J. *J. Phys. Chem. A* **2011**, 115, (5), 577-585.

164. Plaza-Medina, E. F.; Rodriguez-Cordoba, W.; Peon, J. *J. Phys. Chem. A* **2011**, 115, (35), 9782-9789.

165. Ioki, Y. *J. Chem. Soc., Perkin Trans. 2* **1977**, (10), 1240-2.

166. Hamanoue, K.; Amano, M.; Kimoto, M.; Kajiwara, Y.; Nakayama, T.; Teranishi, H. *J. Am. Chem. Soc.* **1984**, 106, (20), 5993-7.

167. Hamanoue, K.; Nakayama, T.; Kajiwara, K.; Yamanaka, S.; Ushida, K. *J. Chem. Soc., Faraday Trans.* **1992**, 88, (21), 3145-51.

168. Fukuhara, K.; Kurihara, M.; Miyata, N. *J. Am. Chem. Soc.* **2001**, 123, (36), 8662-8666.

169. Hamanoue, K.; Nakayama, T.; Amijima, Y.; Ibuki, K. *Chem. Phys. Lett.* **1997**, 267, (1,2), 165-170.

170. Chapman, O. L.; Heckert, D. C.; Reasoner, J. W.; Thackaberry, S. P. *J. Am. Chem. Soc.* **1966**, 88, (23), 5550-4.

171. Carlos E Crespo-Hernández, A. V. R. a. B. S. *Mod Chem appl* **2013**, 1, (3).

172. Crespo-Hernandez, C. E.; Burdzinski, G.; Arce, R. *J. Phys. Chem. A* **2008**, 112, (28), 6313-6319.

173. Garcia-Berrios, Z. I.; Arce, R. *J. Phys. Chem. A* **2012**, 116, (14), 3652-3664.

174. Brown, H. W.; Pimentel, G. C. *J. Chem. Phys.* **1958**, 29, 883-8.

175. Schalk, O.; Yang, J.-P.; Hertwig, A.; Hippler, H.; Unterreiner, A. N. *Mol. Phys.* **2009**, 107, (20), 2159-2167.

176. Suhr Kirketerp, M.-B.; Axman Petersen, M.; Wanko, M.; Andres Espinosa Leal, L.; Zettergren, H.; Raymo, F. M.; Rubio, A.; Broendsted Nielsen, M.; Broendsted Nielsen, S. *ChemPhysChem* **2009**, 10, (8), 1207-1209.

177. Agmon, N. *J. Phys. Chem. A* **2005**, 109, (1), 13-35.

178. Frick, E.; Ernst, H. A.; Voll, D.; Wolf, T. J. A.; Unterreiner, A.-N.; Barner-Kowollik, C. *Polym. Chem.* **2014**, 5, (17), 5053-5068.

179. Yagci, Y.; Jockusch, S.; Turro, N. J. *Macromolecules (Washington, DC, U. S.)* **2010,** 43, (15), 6245-6260.

180. Yilmaz, G.; Aydogan, B.; Temel, G.; Arsu, N.; Moszner, N.; Yagci, Y. *Macromolecules (Washington, DC, U. S.)* **2010,** 43, (10), 4520-4526.

181. Yilmaz, G.; Tuzun, A.; Yagci, Y. *J. Polym. Sci., Part A: Polym. Chem.* **2010,** 48, (22), 5120-5125.

182. Balta, D. K.; Cetiner, N.; Temel, G.; Turgut, Z.; Arsu, N. *J. Photochem. Photobiol., A* **2008,** 199, (2-3), 316-321.

183. Balta, D. K.; Temel, G.; Aydin, M.; Arsu, N. *Eur. Polym. J.* **2010,** 46, (6), 1374-1379.

184. Lalevee, J.; Morlet-Savary, F.; El Roz, M.; Allonas, X.; Fouassier, J. P. *Macromol. Chem. Phys.* **2009,** 210, (5), 311-319.

185. Saraiva, M. F.; Couri, M. R. C.; Le Hyaric, M.; de Almeida, M. V. *Tetrahedron* **2009,** 65, (18), 3563-3572.

186. Dietlin, C.; Allonas, X.; Morlet-Savary, F.; Fouassier, J. P.; Visconti, M.; Norcini, G.; Romagnano, S. *J. Appl. Polym. Sci.* **2008,** 109, (2), 825-833.

187. Lewis, F. D.; Lauterbach, R. T.; Heine, H. G.; Hartmann, W.; Rudolph, H. *J. Am. Chem. Soc.* **1975,** 97, (6), 1519-25.

188. Gruber, H. F. *Prog. Polym. Sci.* **1992,** 17, (6), 953-1044.

189. Hageman, H. J. *Prog. Org. Coat.* **1985,** 13, (2), 123-50.

190. Anseth, K. S.; Newman, S. M.; Bowman, C. N. *Adv. Polym. Sci.* **1995,** 122, (Biopolymers II), 177-217.

191. Ge, J.; Trujillo, M.; Stansbury, J. *Dent. Mater.* **2005,** 21, (12), 1163-1169.

192. Sun, H.-B.; Kawata, S. *Adv. Polym. Sci.* **2004,** 170, (NMR, 3D Analysis, Photopolymerization), 169-273.

193. Gunzler, F.; Wong, E. H. H.; Koo, S. P. S.; Junkers, T.; Barner-Kowollik, C. *Macromolecules (Washington, DC, U. S.)* **2009,** 42, (5), 1488-1493.

194. Voll, D.; Junkers, T.; Barner-Kowollik, C. *Macromolecules (Washington, DC, U. S.)* **2011,** 44, (8), 2542-2551.

195. Wolf, T. J. A.; Voll, D.; Barner-Kowollik, C.; Unterreiner, A.-N. *Macromolecules (Washington, DC, U. S.)* **2012,** 45, (5), 2257-2266.

196. Buback, M. K., A. *Macromol Chem Physic* **2003,** (204), 632-637.

197. Lipson, M.; Turro, N. J. *J. Photochem. Photobiol., A* **1996,** 99, (2-3), 93-96.

198. Morlet-Savary, F.; Allonas, X.; Dietlin, C.; Malval, J. P.; Fouassier, J. P. *J. Photochem. Photobiol., A* **2008,** 197, (2-3), 342-350.

199. Jockusch, S.; Landis, M. S.; Freiermuth, B.; Turro, N. J. *Macromolecules* **2001,** 34, (6), 1619-1626.

200. Seidl, B.; Liska, R.; Grabner, G. *J. Photochem. Photobiol., A* **2006,** 180, (1-2), 109-117.

201. Jockusch, S.; Koptyug, I. V.; McGarry, P. F.; Sluggett, G. W.; Turro, N. J.; Watkins, D. M. *J. Am. Chem. Soc.* **1997,** 119, (47), 11495-11501.

202. Ma, C.; Du, Y.; Kwok, W. M.; Phillips, D. L. *Chem. - Eur. J.* **2007,** 13, (8), 2290-2305.

203. Turro, N. J. *University Science Books* **1991,** Mill Valley, California.

204. Lu, K.-T.; Weinhold, F.; Weisshaar, J. C. *J. Chem. Phys.* **1995,** 102, (17), 6787-805.

205. Spichty, M.; Turro, N. J.; Rist, G.; Birbaum, J. L.; Dietliker, K.; Wolf, J. P.; Gescheidt, G. *J. Photochem. Photobiol., A* **2001,** 142, (2-3), 209-213.

List of Publications

Publications

E. Frick†, H. A. Ernst†, D. Voll, T. J. A. Wolf, A.-N. Unterreiner, C. Barner-Kowollik, Studying the polymerization initiation efficiency of acetophenone-type initiators via PLP-ESI-MS and femtosecond spectroscopy, *Polym. Chem.*, **2014**, *5*, 5053-5068.

† These authors contributed equally to this publication.

H. A. Ernst, T. J. A. Wolf, O. Schalk, N. González-García, A. E. Boguslavskiy, A. Stolow, M. Olzmann, A.-N. Unterreiner, Ultrafast dynamics of *ortho*-nitrophenol: an experimental and theoretical study, *J. Phys. Chem. A*, **2015**, submitted.

Conference contributions

H. A. Ernst, T. J. A. Wolf, N. González-García, A.-N. Unterreiner, M. Olzmann, Femtosecond pump-probe investigation of *ortho*-nitrophenol and its constitutional isomers in solution, *Bunsentagung 2013, 112th General Assembly of the German Bunsen Society for Physical Chemistry*, Karlsruhe, **2013**, Poster.

H. A. Ernst, T. J. A. Wolf, N. González-García, A.-N. Unterreiner, M. Olzmann, Ultrafast dynamics of *ortho*-nitrophenol in solution, *11th International Conference on Femtochemistry (FEMTO 11)*, Copenhagen, Denmark, **2013**, Poster.

H. A. Ernst, E. Frick, D. Voll, T. J. A. Wolf, A.-N. Unterreiner, C. Barner-Kowollik, Investigation of three selected photoinitiators via femtosecond transient absorption spectroscopy, TD-DFT calculations and PLP-ESI-MS, *Bunsentagung 2014, 113th General Assembly of the German Bunsen Society for Physical Chemistry*, Hamburg, **2014**, Poster.

Curriculum Vitae

Personal data

Name:	Hanna Angela Ernst
Date of birth:	February 3, 1987
Place of birth:	Bühl (Baden)
Nationality:	German

School education

1993-1997	Lothar von Kübel Grundschule (Sinzheim)
1997-1999	Heimschule Lender (Sasbach)
1999-2006	Ludwig-Wilhelm-Gymnasium (Rastatt)
2006	Abitur

Academic studies

2006-2011	Studies in chemistry, Karlsruhe Institute of Technology (KIT; formerly University of Karlsruhe (TH))
2008	Intermediate diploma in chemistry, Karlsruhe Institute of Technology (KIT)
09/2011	Diploma in chemistry, Karlsruhe Institute of Technology (KIT)

Since 01/2012	Research assistant at the Institute of Molecular Physical Chemistry, Karlsruhe Institute of Technology (KIT), working group of PD Dr. A.-N. Unterreiner
01/01-12/31/2012	Scholarship by the Karlsruhe School of Optics and Photonics (KSOP)
2014	Scholarship by the Karlsruhe House of Young Scientists (KHYS) for visiting researcher

Danksagung

Mein besonderer Dank gilt Herrn PD Dr. Andreas Unterreiner für die vielseitige und herausfordernde Themenstellung, sowie die enge Betreuung während meiner Promotion.

Für die finanzielle Unterstützung und die ideelle Förderung möchte ich mich bei der Karlsruhe School of Optics and Photonics (KSOP), sowie bei meinen Mentoren Dr. Franco Weth und PD Dr. Oliver Hampe bedanken.

Ebenfalls möchte ich Herrn Prof. Dr. Matthias Olzmann für die Unterstützung und hilfreiche Diskussion im Rahmen des *ortho*-Nitrophenol Projektes danken.

Dank geht auch an C. García Iriepa, T. J. A. Wolf und A. E. Boguslavskiy für die Zusammenarbeit bei den durchgeführten CASSCF-Rechnungen.

Den Mitgliedern des Polymer-Kooperationsprojektes möchte ich für die gute und erfolgreiche Zusammenarbeit danken.

Allen Mitarbeitern der Abteilung für Molekulare Physikalische Chemie möchte ich für die Unterstützung während meiner Promotion, die nette Arbeitskreis-Atmosphäre und die stets gute Stimmung danken. Besonderer Dank geht an die „verbliebenen Femtos" Dipl.-Chem. Caroline Schweigert und Dr. Yu Liang.

Weiterhin möchte ich mich bei allen Werkstatt-Angestellten für die gute Zusammenarbeit und die schnelle Umsetzung von Aufträgen bedanken. Insbesondere möchte ich mich bei Dipl.-Ing. Klaus Stree und Dipl.-Ing. Holger Halberstadt für die erfolgreiche „Operation am offenen Laser" bedanken, ohne die eine Weiterarbeit unmöglich gewesen wäre.

Dank geht auch an meine externen Korrekturleser Bastian Süveges, Lena Ernst, Simone Maciej, Tristan Yarolem und Mario Lukacevic, sowie Desmond Kelly, der mir bei sprachlichen Fragestellungen hilfreich zur Seite stand.

Ich danke Bastian für seine unermüdliche Unterstützung und seine stetige Motivation, ohne die diese Arbeit nicht zustande gekommen wäre.

Meiner Familie möchte ich für die Unterstützung während meiner gesamten Studienzeit danken.

Den Mitgliedern der Freiwilligen Feuerwehr Stadt Karlsruhe Abteilung ABC-Zug möchte ich recht herzlich für die gute Kameradschaft, die vielen gemeinsamen Übungsstunden und die überaus humorvolle Atmosphäre danken. Die Übungsabende waren eine stets willkommene Abwechslung.